COMETS

The Earth series traces the historical significance and cultural history of natural phenomena. Written by experts who are passionate about their subject, titles in the series bring together science, art, literature, mythology, religion and popular culture, exploring and explaining the planet we inhabit in new and exciting ways.

Series editor: Daniel Allen

In the same series

Air Peter Adey
Cave Ralph Crane and Lisa Fletcher
Clouds Richard Hamblyn
Comets P. Andrew Karam
Desert Roslynn D. Haynes
Earthquake Andrew Robinson
Fire Stephen J. Pyne
Flood John Withington
Gold Rebecca Zorach
    and Michael W. Phillips Jr
Islands Stephen A. Royle
Lightning Derek M. Elsom

Meteorite Maria Golia
Moon Edgar Williams
Mountain Veronica della Dora
Silver Lindsay Shen
South Pole Elizabeth Leane
Storm John Withington
Swamp Anthony Wilson
Tsunami Richard Hamblyn
Volcano James Hamilton
Water Veronica Strang
Waterfall Brian J. Hudson

Comets

P. Andrew Karam

REAKTION BOOKS

To my two favourite teachers: Father Norman Dickson, S. J.,
and Professor Mark Auburn

Published by Reaktion Books Ltd
Unit 32, Waterside
44–48 Wharf Road
London N1 7UX, UK
www.reaktionbooks.co.uk

First published 2017
Copyright © P. Andrew Karam 2017

Printed and bound in China by Toppan Leefung Printing Limited

A catalogue record for this book is available from the British Library

ISBN 978 1 78023 830 2

CONTENTS

Introduction

Consider how our ancestors viewed the sky – filled with mysterious points of light. Most of these are fixed in place, rising and setting through the night and with the seasons. A very few – the planets – move against the backdrop of stars, but they remain confined to a narrow swathe of the sky, what we today call the ecliptic (or the zodiac). And then there are a few other objects that are different from anything else seen in the sky; they can appear anywhere in the sky, they move against the background stars, and most intriguingly, they change their appearance with time, becoming fuzzy and then sprouting long tails. Such 'stars' are beyond anything our ancestors had seen or imagined or even heard of: in an era before science, they were most likely to think that these objects had been sent by the gods, but they did not know why the gods had chosen to send this particular object or what it was supposed to mean.

This book is about these 'stars' – what we now call comets.

Many millennia have passed since our early ancestors first wondered about these odd objects, and in those years, we have learned a great deal. We now know that comets are among the most ancient bodies in our Solar System and that they are indeed messengers from that time to our own. We have studied them through our telescopes and have even sent spacecraft to rendezvous with them and to bring back samples of their substance.

Of course, our understanding of comets did not spring full-blown from ignorance and myth; it developed and evolved

Comet Lovejoy.

gradually over the centuries. Edmund Halley recognized that there was some predictability about comets and that some – including the one eventually named after him – can return time and again. Using this information, Isaac Newton was able to show that the laws of motion he had developed applied to comets as well as to lesser (and greater) objects – Newton's seventeenth- and eighteenth-century discoveries not only helped elucidate the workings of the universe, but they revolutionized how we see ourselves as a part of the universe. For the first time, we recognized that the universe is something that can be comprehended. With ensuing centuries, we have learned far more about comets, but nothing has had a greater impact on humanity than the realization that we could understand our world and the universe in which it resides.

There is much more to the impact of comets upon our culture than our scientific discoveries: as the most spectacular objects in our skies, comets have impressed artists from prehistory until the present. Cave paintings and petroglyphs show that comets made an impression on the most primitive artists; the comet woven into the Bayeux Tapestry is evidence of their continuing impact on the artists of the Middle Ages; and the repeated use of comets in paintings of the Renaissance and later eras tells us that they continued to impress artists through the ages. Whether for their beauty, for their symbolism or simply for historical accuracy, comets have appeared in artwork for millennia.

Comets have been written into our literature for almost as long as humans have been writing, from ancient Rome through Shakespeare and into the present day. Here, too, comets have been used for historical accuracy, dramatic effect or for their symbolism. In the last century, they have also been used literally, as the basis for short science fiction stories and entire novels.

Comets make an appearance in popular culture on a regular basis. These appearances, incidentally, go beyond science fiction movies and television shows; comets have appeared in popular television shows, in movies and even in video games. Add to this the appearance of comets in science fiction and it is clear that comets have more than a foothold in the popular imagination.

When Julius Caesar was assassinated in 44 BCE, the Roman Empire was on the brink of civil war: within a decade, the empire was relatively stable and under the control of Caesar's adopted son. Part of the reason for this was not only that Caesar was declared a deity by the senate of Rome, but the appearance of a great comet in the skies confirmed his deification in the eyes of the empire's citizens. Appearing in the heavens, comets have inspired religious awe as well as religious dread, and these feelings are not limited to antiquity. As recently as 1997, the Heaven's Gate religious cult committed mass suicide, driven in part by their conviction that the great comet of that year (Hale–Bopp) heralded the end of the world and their chance to move on to a higher level of being. We will examine how comets have been viewed by the world's religions over the ages – how they have inspired hopes and fears, and how they have fitted into (or not) the prophecies of the day.

Comets come and go, but some comets are so bright, so brilliant and so beautiful that they have been labelled 'great'. Halley's Comet is the best known of the 'great comets', but it is hardly the only one. Aristotle was inspired by a great comet in the fourth century BCE, and we have already mentioned the great comet that signalled Caesar's deification. Other comets have grazed the Sun, have been so bright as to appear in broad daylight and have inspired the imagination – in every generation or two, a comet appears in the skies that fires the imagination more than most. These great comets have a disproportionate impact on us: they may not have changed history, but they have certainly played a part in history. A number of these great comets will be discussed along with the impact they had on the people of the time.

After all is said and done, there is more to comets than their appearance in the skies, their appearance in our culture and our scientific understanding of them. Comets have quite literally brought both life and death to the Earth. It is very possible that much of the Earth's water arrived with cometary bombardment early in the history of the Solar System, and scientists have identified complex chemicals in comets – chemicals that might well have helped spark the formation of life on Earth. At the same time, when Comet Shoemaker–Levy 9 fragmented and slammed into the atmosphere of Jupiter, it became obvious that comets are just as likely to bring destruction, especially to a planet that is already home to a living ecosystem. Craters across the Earth, Mars, the Moon and the moons of the outer planets attest to the destructive power of cosmic impacts, and even something as seemingly innocuous as the passage of Earth through comets' tails has raised fears. The role of comets in the creation and the destruction of life will be the subject of the final chapter of this book.

1 The Science of Comets

This is a golden age in comet science. Until recently our study of comets has been almost entirely hands off – photos from Earth or from orbital telescopes – but we are now able to observe and sample up close. Since the first years of the twenty-first century, scientists have directly sampled the tails of comets; sent spacecraft to rendezvous with comets; slammed impactors into comets to study their substance and structure; and on 12 November 2014 even landed a probe on a comet for the first time. Some of what scientists are learning is brand new information, some discoveries have helped to confirm what we learned from afar and some they are still trying to figure out. All of the data tells us that comets are some of the most fascinating objects in the Solar System; but first, let us discuss where comets come from and what they are made of.

When you look at the Solar System you see a lot of rocky planets, plus the asteroids, nestled in close to the Sun; further out the planets, moons and other objects tend to be gas and ice. Saturn's rings, a number of major moons and most of the bodies beyond Neptune are made of ice – in this case, ice includes not just frozen water, but frozen carbon dioxide, frozen methane and a number of other gases cooled to cryogenic temperatures. The Solar System is, in effect, sorted out by temperature: the Sun boiled off many of these volatile materials from the inner Solar System, leaving behind mostly rock; temperatures in the outer Solar System were lower and the gas and ice remained.

Somewhere between five and six billion years ago, there was no Solar System. In our part of the Milky Way, there was only a cloud of gas and dust spinning slowly in space. Even today our galaxy has thousands of dust and gas clouds lacing its structure; some of these clouds are as old as our galaxy and nearly as old as the universe itself. Left on its own, our gas and dust cloud would have remained quietly occupying its own corner of the galaxy. But it was not left on its own – about five or six billion years ago (give or take a little), a nearby star exploded, and when the shockwave ploughed through space, it compressed some of the gas and dust ever so slightly, creating regions that were fractionally denser than others; the denser areas contained more mass and exerted slightly more gravitational force than their more rarified neighbours. Over time an increasing amount of gas and dust collected in these denser areas, increasing their gravity even more. In a handful of millions of years, the cloud had collapsed, forming a nascent star surrounded by a swarm of planets. Minor planets collided and merged, eventually settling out into the Solar System we see today: a few rocky planets warmed by the Sun, a few chilly gas giants and a host of rubble (the asteroids) in between.

Nothing in the universe seems to stand entirely still, and the solar nebula (as the gas and dust cloud is known) was no exception – it was rotating slowly in space. As it collapsed, it spun up, rotating ever faster the smaller it got to conserve angular momentum (the classic example of this sort of thing is an ice skater who spins faster as she pulls in her arms). But there's more to it than that – spinning objects experience centrifugal force. So picture a big cloud of gas that is spinning in space: the rotational velocity will be highest in the middle (equivalent to the equator on Earth) and will be lowest at the top and bottom (the 'poles'). The Earth is solid so this centrifugal force only creates a slight bulge round the equator; a cloud of gas is something else entirely. There comes a point at which the inward pull of gravity is balanced by the outward centrifugal force; when the cloud gets to this point, it will continue to collapse at its poles, but the equator maintains about the same radius. This balance of forces means that the inner part of the cloud becomes a disc. This is why our galaxy is disc-shaped and why all of the major planets in the Solar System are roughly aligned with the Sun's equator in a plane called the ecliptic. But comets are born far outside where the planets reside – in the frigid outer reaches of the Solar System, the centrifugal force is minuscule so there is no longer a disc; comets are knots of material scattered throughout the sky in a sphere centred on the Sun.[1]

These knots of material – 'dirty snowballs' in the words of the American astronomer Fred Whipple (1906–2004)[2] – are generally stuck in the outskirts of the Solar System. Tethered only tenuously by the Sun's gravity and warmed not at all, these lumps of ice and dust circle the Sun more slowly than any of the closer objects – an object loitering in the outermost Solar System moves at the speed of a brisk walk. Objects this far out are also easily perturbed by passing stars: although most of the bodies in the Solar System had settled out into the ecliptic, repeated tugging by the stars and gravitational scattering by Jupiter and Saturn pulled the furthest ones into more random orbits until they formed a spherical cloud that extended outwards halfway to the nearest stars. Every now and again one of them will be perturbed even more and it will plunge in towards the Sun. As they warm up, the most volatile gases boil off, carrying dust with them and forming spectacular tails. These are the comets and they are among the most ancient objects circling the Sun.

The structure and substance of comets

Comets have a distinct structure: when they are in the outer reaches of the Solar System, there is the comet itself; as it draws closer to the Sun, the outer layers of ice boil off and the comet grows one or more tails along with an envelope of gas that is loosely bound by the comet's feeble gravity. When we look at a comet in our skies, the tiniest part of what we see is what contains the most material, the nucleus. The outer layers (coma and tails) come and go but the nucleus is the heart of the comet. We can have a comet without the tails, but without the nucleus there is no comet.

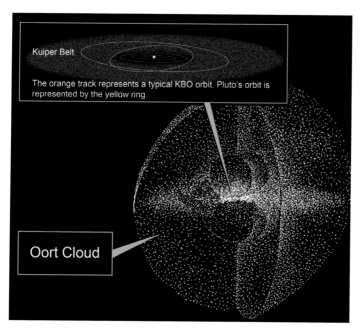

Kuiper Belt

The orange track represents a typical KBO orbit. Pluto's orbit is represented by the yellow ring.

Oort Cloud

The Oort Cloud, which contains billions or trillions of comets, and dwarfs the roughly 500 million km radius of the inner Solar System.

As long as the comet is out where it normally resides, far away from the Sun in the Oort Cloud – named for the Dutch astronomer Jan Oort, who first proposed its existence – or in the Kuiper Belt near Neptune, comets are nothing more than chunks of ice mixed with dust and rocks. Comets have also been described as piles of rubble – masses of ice and rock, although more ice than rock, clumped together and loosely held in place by the objects' weak gravity. Add a sprinkling of interplanetary and interstellar dust to the surface, along with the odd cluster of other molecules, and you have Whipple's dirty snowball. Whether we call them piles of rubble or snowballs is almost immaterial; the important thing to remember is that they are small – the nucleus is usually only about 10 km (6 miles) in diameter – and composed mostly of ice with a smattering of other material, such as dust, rock and some complex molecules. The molecules can include simple gases (such as methane, carbon dioxide and cyanogen), as well as more complicated molecules such as amino acids, the molecular building blocks from which proteins are manufactured by our cells. These molecules – both simple and complex – form in the space between the planets and stars; there are some scientists who believe that these might have helped seed the early Earth with the ingredients from which life eventually formed.[3]

The Earth is differentiated: there is a core made of iron and nickel, a mantle made of iron-rich minerals and a crust composed of minerals rich in silica. In fact, all of the major bodies in the Solar System are differentiated; minor objects are not. The deciding factors are size and chemical composition. The same supernova whose shock wave initiated the collapse of the solar nebula also produced uranium, thorium and even more exotic radioactive elements; these give off heat when they decay, and whatever object contains them will warm up. The larger an object is, the more radioactivity it will contain and the more heat it will retain – smaller objects generate only traces of heat and that heat radiates rapidly into space.

Size has another effect as well – it affects an object's shape. Larger objects have higher gravity, and this gravity will pull them into a spherical shape; even the Earth, with all its mountains and deep-sea trenches is (relatively speaking) more spherical than a ball bearing.

Comets are, by Solar System standards, tiny – the nucleus is usually no more than a few tens of kilometres in diameter. They are too small to be pulled into a spherical shape, too small to

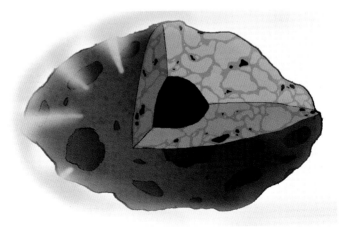

The structure of
a comet.

generate much in the way of radioactive heat (and far too small to retain it) and too small to become differentiated. What this means is that comets are fairly simple: with no internal structure to speak of, they are merely piles of ice (including frozen gases), dust and rock, hence Whipple's 'dirty snowball'. In the outer Solar System, that is about all there is; but as comets approach the Sun, things start to get more interesting.

Substance	Boiling temperature (°c)		Distance from Sun (au)*	Nearest planet
	sea level	vacuum		
Water	100	-68	1.4	Mars
Ammonia	-33	-100	1.6	(1.5 au)
Carbon dioxide	-79	-130	2.9	Asteroid belt (2.7 au)
Methane	-164	-202	10.5	Saturn (9.5 au)
Carbon monoxide	-192	-213	19	Neptune (19 au)

*au (astronomical unit) is the average distance from Earth to Sun, or about 150 million km.

Every substance has a temperature at which it turns to vapour – its boiling point – and this includes the substances of which comets are made. The table here summarizes the boiling temperatures for a number of these substances as well as the distance from the Sun at which these temperatures will be encountered. This helps to explain why comets lack a tail in the outer Solar System and why their comas and tails form as they approach the Sun. Water, for example, boils at 100°c (212°f) at sea level, but as low as -68°c (-90°f) in the vacuum of space. At atmospheric pressure, ammonia vaporizes at a temperature of about -33°c (-27°f), but in a vacuum that temperature drops to about -100°c. The other volatile substances of which comets are

The nucleus of
Comet Tempel 1,
photographed by
the Deep Impact
probe prior to impact
on 15 September 2005.

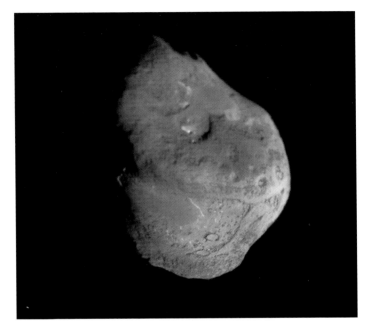

composed each have their own vaporization temperatures: as the comet falls into the inner Solar System, it warms, and the substances boil off into space, carrying with them some of the dust and other molecules that coat the comet's surface and that are mixed into the ice. As the comet passes the orbit of Neptune, the carbon monoxide begins to boil and, around Saturn, it starts to off-gas methane – but these make up less than 10 per cent of the comet. By the time the comet passes the asteroid belt, it is starting to give off carbon dioxide, and around Mars, both water and ammonia are boiling off into space.

Significantly, the material does not evaporate uniformly across the surface of the comet, and the gas comes out in jets. Some places, for example, will be shaded, while others are darker-coloured and absorb more solar energy. Some areas might be paved with gravel or dust; some might be covered with materials that have a lower boiling temperature; some locations might simply be stronger or weaker than others. What this means is that some areas are more likely to evaporate first. As the ice melts and boils, it starts to form a crater, and as this crater deepens, it becomes almost like a rocket nozzle, focusing the gas into a narrower jet.[4]

The gases that boil off the comet form some distinct features. One is the coma, a tenuous envelope of gas and dust that surrounds the comet and that gives it a characteristically fuzzy appearance. In the chilly outer Solar System, the coma might only be the size of the Earth; closer to the Sun it can swell to the size of Jupiter or more. Warmer temperatures mean more out-gassing, and more out-gassing means more gas and dust surrounding the comet – in 2007 Comet Holmes had an outburst of gas and dust that made it temporarily larger than the Sun.

The coma changes size as the comet approaches the Sun, but not in the way that one might expect. As mentioned earlier, the coma will grow in size as the comet warms up, but the solar wind (the flow of gas outwards from the Sun) strengthens as well. By the time the comet reaches the orbit

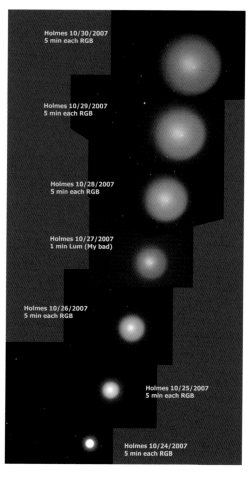

Photo montage of Comet 17P/ Holmes following its spectacular outburst on 24 October 2007.

of Mars, the solar wind is starting to strip off the outer layers, and the coma begins to shrink again.

The coma can be impressive, but the tail is what makes comets so spectacular. A comet's nucleus might be no more than a few tens of kilometres in diameter, and the coma can be more than 1 million km (620,000 miles) across. A comet's tail can extend over 500 million km, stretching halfway across the sky. But of course there's more to the comet's tail (literally and figuratively) than meets the eye, starting with the fact that most comets have more than one tail – there might be tails made of dust, gas and/or ions, and they all behave differently. Finally, the tails do not necessarily trail behind the comet as it travels through space; they almost invariably point away from the Sun, following the comet as it approaches the Sun and preceding it on the return to the outer Solar System.

The most obvious of the tails is the dust tail – this is the dust blown out of the comet by the escaping jets of gas. Blowing out of the Sun is a constant stream of gas and charged particles (protons, electrons and helium atoms); it moves through space at speeds of several hundred km per second. Just as a stiff breeze will carry dust with it, the solar wind carries with it the dust particles. But there is another propulsive force as well – light itself can exert a very small pressure on dust in space, and the pressure of sunlight also helps to push cometary dust away from the Sun.

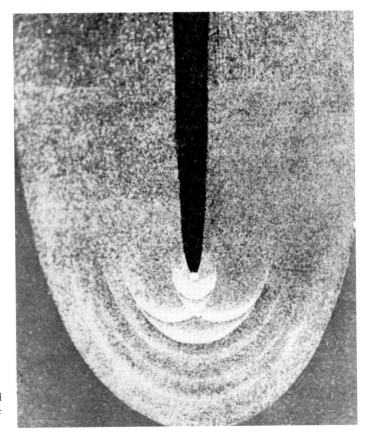

Coggia's Comet (July 1864), illustrating how interactions between the comet, heat from the Sun and the solar wind can cause layered envelopes of gas in the comet's coma.

The light travels directly outwards from the Sun while the solar wind follows the magnetic field lines that lace the Solar System. These two processes can blow two streams of material in slightly different directions, creating more than one tail. Once in space, the dust scatters and reflects the sunlight, making it visible throughout the Solar System. The dust tail tends to be curved, roughly tracing out the path of the comet's orbit through space.

    Gas, of course, is also emitted by the comet; the gas is what carried the dust into space. The gas is also carried outwards by

Comet Hale–Bopp, photograph showing the dust and ion tails.

the solar wind, although, being largely transparent, it tends to be unaffected by light pressure. This gas can also be ionized (more on this in a moment), forming a plasma similar to what is found inside fluorescent lights. Like a fluorescent lamp, the plasma emits light, the exact colour of which depends on exactly which atoms and molecules are present.

Ultraviolet light from the Sun is powerful enough to strip electrons off the gas (this process is called ionization), and this can create still another tail called an ion tail. The gas and ion tails will generally be blown directly away from the Sun, giving them a different direction and shape than the dust tail and, on a large scale, the ion tail (being electrically charged) will be shaped by the Sun's magnetic field. The ions are electrically charged, forming a plasma similar to what is found in a neon or fluorescent light. This plasma is exposed to the solar magnetic field – charged particles passing through a magnetic field will generate their own magnetic field. So, along with the Sun, the planets and some of the larger moons and asteroids, comets can have their own magnetic field – at least, when they are close enough to the Sun to expel gas that can become ionized.

In addition to the formation of the tails via gas and dust expulsion, comets lose larger chunks as well; these gas jets are not 'clean', and they litter space with traces of cometary debris. When this debris enters our atmosphere, we see it as meteors. Dust, sand-sized mineral grains and even small pebbles are expelled from the comet, and this debris is going to follow its own path around the Sun. In general, this path will be close to that of the comet, but smeared out somewhat since the jets will be blown off in all directions. Some cometary orbits intersect that of the Earth, and the Earth passes through the orbits every year. When it does so, some of this debris is captured by the Earth's gravitational field, raining down into our atmosphere as meteors. For example, the annual Perseid meteor shower is associated with debris expelled from Comet Swift–Tuttle; the Ursid shower is associated with comet 8P/Tuttle; and the Orinid meteor shower seems to originate from debris that has been blown off Comet Halley.

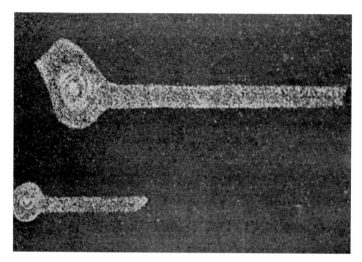

Comet Biela (September 1852) apparently broke into two major pieces during one pass by the Sun, appearing as a double comet during its future appearances.

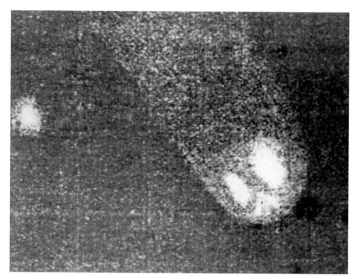

The double Olinda Comet (February 1860) likely also started as a single comet before thermal and gravitational stresses caused it to break into two large pieces.

Incidentally, dust and debris are not the only cometary material swept up by the Earth. During the 1910 appearance of Comet Halley, the Earth passed through the comet's tail. Among the gases identified in the tail was the toxic gas cyanogen. This occasioned some worry among a few members of the public. What they did not know is that, while cyanogen was indeed present, it was in such trace amounts that the health

Gas jets emanating from Comet Hartley, obtained during a flyby during the NASA EPOXI mission.

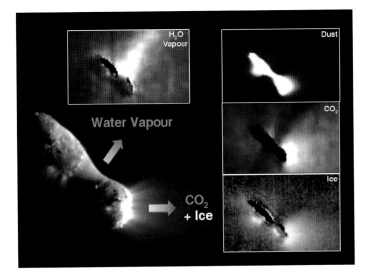

impact was non-existent. And, in fact, this is something that not many would guess – not just gas, but for all its blazing glory, the gas and dust in a comet's tail is so rarified as to be scarcely more than vacuum itself.

How they move

As with every other object in the universe, comets' motions are ruled by gravity; being part of the Solar System, the primary influence on their motion is that of the Sun. That being said, in the Oort Cloud, where most of the comets reside, the tug of an occasional passing star can be enough to nudge a comet into the inner Solar System. As the comet drops in past the giant planets – especially Jupiter – it can be pulled even further off course; it can be captured into orbit around Jupiter, can be flung into a different solar orbit or can even be ejected from the Solar System.

Although all the planets and most of the asteroids lie in the plane of the ecliptic, most comets were born beyond the region where the solar nebula collapsed into a disc. In the Oort Cloud, the comets buzz around the Sun like a swarm of bees in a roughly spherical shell that begins about 5,000 times as far from

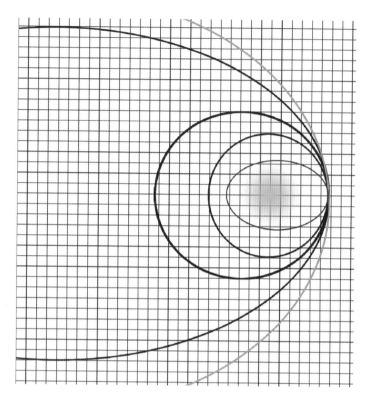

Conic sections and examples of cometary orbits.

the Sun as the Earth is and extends almost a quarter of the way to the nearest star. This means that the comets that are nudged out of their regular orbits towards the Sun can come from any part of the sky. Some come in along the ecliptic, while others plunge sunwards almost perpendicularly, and any angle in between these extremes. So compared with the bodies of the inner Solar System, the fact that comets are not confined to the ecliptic makes them unusual.

The other thing that makes comets' orbits different from what we are used to in the inner parts of the Solar System is that their orbits are far from circular – the orbit of the planets (with the exception of the minor planet Pluto) deviate only very slightly from being perfect circles, but cometary orbits are incredibly varied. Comets in the Oort Cloud might actually have nearly circular orbits, although nobody is sure about this because we cannot see such small objects at so great a distance. However, when comets are perturbed towards the Sun, their orbits change considerably. Eccentricity (as it applies to orbits) is a measure of how far an orbit deviates from perfect circularity; an orbit that is perfectly circular has an eccentricity of zero, and the more elongated and narrower the orbit the higher the number. Any ellipse will have an eccentricity of between zero and one; any object in an orbit like this will cycle endlessly unless something comes along to yank it into a different orbit.

But not all orbits are closed like this, and there are other possibilities – not all comets return. A comet's path might be parabolic or hyperbolic, plunging into the inner Solar System and whipping past the Sun, never to return. Incidentally, even comets that are in elliptical orbits can be thrown into non-returning orbits if they pass close enough to Jupiter – in the early days of the Solar System, it is thought that Jupiter, with its tremendous mass and gravity, helped to protect the planets of the inner Solar System by flinging untold numbers of comets and asteroids into the System's outer reaches, diverting them into non-returning orbits and ejecting them out into interstellar space. Not only that, but Jupiter can capture comets as well – for instance, Comet Shoemaker–Levy 9 had already been captured by Jupiter's gravity when it was discovered in 1993; it was ripped

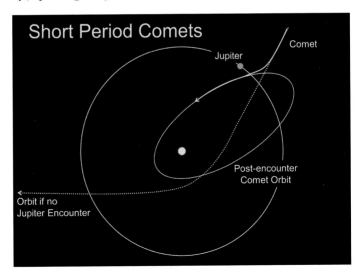

Short Period Comets

Comet

Jupiter

Post-encounter
Comet Orbit

Orbit if no
Jupiter Encounter

An example of a non-returning comet being captured by Jupiter's gravity.

apart by tidal forces and smashed into Jupiter over the course of a week in mid-July 1994.

So comets' orbits are not only different from those of the planets and asteroids, but they can change dramatically over time as well. That is not all that varies: the speed of any object in orbit around the Sun depends primarily on its distance and on the shape of the orbit; the further from the Sun an object lies, the more slowly it moves. As far as the effect of distance, orbital velocities fall off with increasing distance – Mercury zips along in its orbit at nearly 50 km/sec, while Neptune dawdles along at less than 6 km/sec. Further out, at Oort Cloud distances from the Sun, the orbital speeds are even lower: at 100,000 AU, a comet in a nearly circular orbit will move at a speed of only about one-tenth of a km/sec – faster than most cars, but slower than most aeroplanes. Another factor is the shape of an orbit; highly eccentric orbits tend to have the slowest velocities at their greatest distance, where they pause briefly before falling back towards the Sun, but they are among the fastest-moving objects in the Solar System at their closest approach. Putting all this together, comets have the greatest variety of orbits of any objects in the Solar System – they can range from nearly circular to highly eccentric non-returning hyperbolae – and the speed of the same comet can vary by two or more orders of magnitude as it moves from the Oort Cloud or the Kuiper Belt to its nearest approach to the Sun.

Another factor about comets' orbits pertains to their visibility on Earth. All of the planets and objects in the asteroid belt lie very close to the plane of the Sun's equator – the ecliptic. Comets – especially Oort Cloud comets – do not. They can plunge into the inner portion of the Solar System from any angle. So consider a comet that enters the inner Solar System from the direction of the Sun's north pole (which is only about 23.5° away from the Earth's north pole): for most of its passage through the Solar System, this comet will be visible, if at all, only from the parts of Earth that can see that part of the sky. Someone observing from, say, Australia or most of South America will not be able to see the comet because the Earth will be in the way;

observers in the northern hemisphere, on the other hand, will be able to see the comet at its faintest and can watch it grow in brilliance as it approaches the Sun and sprouts its tail. As the comet approaches the Sun, it will pass through the plane of the ecliptic. At this point, it can be seen from the southern hemisphere but not from North America or Europe; it will remain invisible to observers in those continents until its return journey, when it again passes into northern skies. Thus, northern observers will see the comet on the inbound and outbound legs but might miss it at its most brilliant, just the opposite of what observers south of the equator will see. Depending on the comet's actual orbit, it might be too dim to see at all when in northern skies. In this case, only southern viewers would be aware of it.

How we know what we know about comets

For millennia the only information people had about comets came from naked-eye observations; with the eye, there just was not much that could be learned aside from its motion across the skies. The invention of the telescope made it possible to learn more. With precise observations of a comet out to greater distances, the astronomers of the day could start to determine orbits and could map the distribution of comets in space and could even start to determine their size and mass. Eventually scientists developed additional instruments that made it possible to learn much more, including their chemical composition. This, plus ever-more-powerful telescopes, gave astronomers even more information about comets; they were able to start making much more educated guesses about what comets were made of and where they came from. Finally, with the return of Comet Halley in 1986, we started launching spacecraft to look at comets from close range. These projects culminated in later missions that slammed comets with impactors, and then we successfully landed a probe on a comet in 2014. We can examine these various methods one at a time.

The *naked eye* was our first instrument and, sad to say, it has some limitations. It can only see a narrow slice of the electromagnetic spectrum, for example: the universe is awash in radio waves, x-rays, infrared, ultraviolet and gamma radiation, none of which our eyes can detect. One astronomer told an undergraduate class he was teaching, 'Nothing interesting happens in the visible wavelengths,' and he has a point.[5] Our eyes are also not very good at detecting very dim objects; they can only see objects that are large enough to be resolved; and they are not hooked up to a tremendously reliable data storage system. But until the telescope was invented in the early years of the seventeenth century, the eye was the only instrument available; what the early astronomers were able to accomplish in spite of the eye's limitations is impressive.

So we have seen what the eye will *not* do; what about what the eye *can* do? One thing it can do is to observe relative locations and relative brightness. A trained observer can notice that an object is moving against the background of fixed stars and can also notice that an object is waxing or waning in brightness. The relative position shows how an object is moving across the sky; the brightness tells us whether an object is moving closer or farther away. This means that a trained observer can make observations that can help determine a comet's orbit. At the very least, naked-eye observations can make it possible to predict where a comet will be day to day (or from year to year). With better mathematical tools, this information can also be used to help calculate a fairly precise orbit of the comet. Tycho Brahe (1546–1601) was the last of the great naked-eye astronomers. His observations set the stage for the telescopic observations of the following century.

Telescopes were the next step, and telescopic observations did a lot more than help seventeenth-century astronomers determine cometary orbits (but more on that later). Telescopes give us two advantages over the naked eye: they collect more light than our eyes can, and they magnify objects. The first attribute is both more important and easier to explain. The pupil of our eye is, at best, only 1 cm in diameter; a telescope mirror or lens with a diameter of, say, 5 cm is 25 times bigger in area and can collect 25 times as much light as can a single eye; a 10-cm telescope can

Tycho Brahe, the last of the great naked-eye astronomers, at his Uraniborg observatory. Artist unknown, hand-coloured engraving from the 1598 printing of Brahe's *Astronomiae instauratae mechanica.*

collect 100 times as much light as a single eye (and 50 times as much light as two eyes). It is impressive, but consider that the European Southern Observatory (ESO) has *four* telescopes, each of which is holding an 8.2-m mirror. This observatory can collect more than a million times as much light as a naked-eye observer and can see objects that are more than a million times dimmer than anything a person can see. In reality it can collect even more light, since a telescope can be trained on an object for hours at a time; telescopes in space can be trained on the same patch of sky for days or weeks to collect even more light. Since brightness drops off with the inverse square of distance, this means that the ESO telescopes can see objects that are a minimum of a thousand times as far away as the naked eye can. When you also consider the length of time it can observe, this ratio goes up even further (collecting light for an hour to make a single image, for example, increases this factor by a further factor of 3,600). By increasing the light collection area and the light collection time, telescopes can help us to see dimmer objects than is possible with the naked eye and can let us see any object to a far greater distance.

The second attribute is somewhat more difficult to explain, partly because the term 'magnify' is rather imprecise. More precisely put, a telescope has greater resolving power than does the human eye and has a smaller field of view. This means that, instead of, say, the planet Jupiter being just a small dot of light in a wide field of view, it becomes a disc that fills a very small field of view. The greater resolving power means that, on this disc, a telescope can resolve (can 'see') smaller objects than can the eye. The net result is that with a telescope, we can see not only things that are far dimmer, but that are far smaller than what the eye will reveal.

In addition, there is one more attribute: instruments that cannot be hooked up to a naked eye can be hooked up to a telescope. Not only can a telescope collect light from a tiny part of the sky for hours or days at a time, but it can make it possible to analyse that light in any of a number of ways. Breaking the light down into its component colours can tell us a lot – hydrogen, for example, will glow a very specific shade of red under some circumstances, while oxygen will shine green and sodium will glow yellow; by observing an object through precise filters, or by carefully observing how colours from an object are distributed (called a spectrum), astronomers can tell something about its chemical composition. They can do this for distant stars, and they can do it for the far nearer comets.

Instruments can do more than simply look at spectra; they can also analyse light of any colour; and they can extend our senses beyond what we can see unassisted. Take the first of these – even something as simple as precisely measuring the brightness of an object every minute can tell us a lot. Consider the Earth, for example: the land tends to be green or brown, while the oceans are blue. If we can measure the brightness of blue and green from minute to minute, we can tell whether or not the half of the Earth we are looking at is mostly land or water. If we look even more carefully, we can even tell the difference between when the Pacific Ocean is passing through our field of view compared to the Atlantic, or Asia versus the Americas. Over time, we can not only tell how long it takes the Earth to rotate on its axis, but we can start to tell (roughly speaking) how much of the surface is composed of water as opposed to land; if we analyse for white as well, we can even assess the amount of cloud cover and ice. Extending this simple brightness analysis to comets can help us to understand how quickly they rotate, how much variability there is in the colour (and composition) of the comet's surface and (when they approach the Sun) which gases are boiling off the surface and in what concentrations.

Finally, our instruments can extend our senses into realms of the spectrum far beyond what nature gave us. Our eyes, good as they are, can only detect a narrow slice of the electromagnetic spectrum – what we call visible light – that ranges from about 400 to 700 nanometers (nm); 400 nm is violet and 700 nm is red light. Anything outside this range – infrared, ultraviolet, radio waves, x-rays and gamma rays – is invisible to our eyes. But instruments can bring all of those bands, ranging in wavelength from a tiny fraction of a nanometer up through billions of nanometers, into view for us, and each band gives us different information. Although comets do not emit much in the way of gamma rays or x-rays, astronomers can learn quite a bit from thermal, ultraviolet, radio, millimetre and other longer wavelengths. By giving us the ability to expand the range of wavelengths over which we can gather information, our instruments give us the ability to learn far more than is apparent in visible light alone.

Spacecraft are the final set of tools from which we have learned so much of what we now know about comets. After

Two of the four
mammoth telescopes
at the European
Southern Observatory
.(ESO) high in the
Andes Mountains
of Chile.

studying them from afar through our telescopes, we finally sent a flotilla of craft to rendezvous with Halley's Comet during its return to the inner Solar System in 1986. For weeks these craft kept formation with the comet, studying it with their cameras (in multiple wavelengths) and imaging the comet at distances ranging from 1,000 to 16,000 km (620–10,000 miles). These observations were the first that ever directly imaged a comet's nucleus and were the first to confirm that Whipple's 'dirty snowball' model was actually a fairly good description of these objects.

Since that rendezvous, other missions have collected samples of cometary dust and gas (including returning some samples to

Photograph from the surface of Comet Churyumov-Gerasimenko, taken by the Rosetta space probe shortly after becoming the first spacecraft to land on a comet on 6 August 2014.

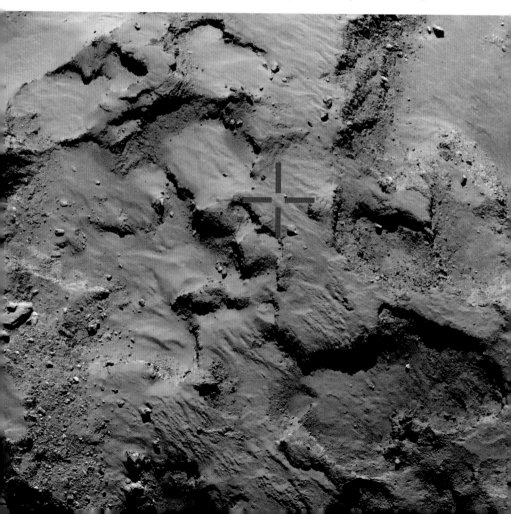

Earth for laboratory analysis), fired a projectile into a comet and even landed a small probe onto Comet Churyumov–Gerasimenko in August 2014. This is why the last decade or so can be considered a golden age in cometary science. For the first time in history, we have not only captured close-range photos that show us exactly what multiple comets look like, but we have even collected samples of their very essence. While there is still a tremendous amount to learn – each question answered seems to spawn another dozen or so – we seem to have a good idea of what comets are made of and how they are put together. But it was not always this way. In the next chapter, we will see how our understanding of comets has changed over the centuries.

Example of spectra showing absorption lines from several stars.

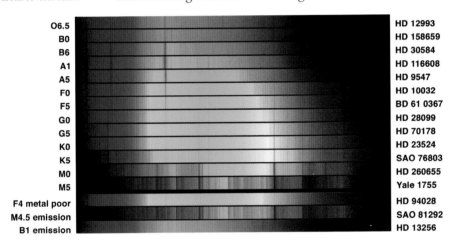

O6.5		HD 12993
B0		HD 158659
B6		HD 30584
A1		HD 116608
A5		HD 9547
F0		HD 10032
F5		BD 61 0367
G0		HD 28099
G5		HD 70178
K0		HD 23524
K5		SAO 76803
M0		HD 260655
M5		Yale 1755
F4 metal poor		HD 94028
M4.5 emission		SAO 81292
B1 emission		HD 13256

2 Studying Comets through the Ages

In the ancient world, comets were imbued with far more significance than they enjoy today – they were the heralds of change and disaster or messages from the gods (even if their meaning was sometimes less than clear). Comets gave rise to myths and legends; they inspired art ranging from cave paintings to engravings to mosaics. What we know (or think that we know) about comets has changed over the centuries as our scientific instruments and insights have evolved. Through the centuries, scientific thinking emerged as a way to understand nature, including those parts of nature that lay beyond direct human reach – Tycho Brahe, for example, showed that comets lay beyond the atmosphere; Johannes Kepler and Isaac Newton developed mathematical explanations of comets' movements through the skies; and Edmund Halley recognized that comets could, and did, return in a predictable fashion. As physicists and astronomers began to make successful predictions of these celestial phenomena, they (and the rest of humanity) began to appreciate that our world and the universe in which it resides follow specific physical laws. This realization itself had a profound impact not only on science, but on culture.

Pre-scientific era

The scientific era is only a handful of centuries old; before it arrived, the only tools available for observations were the naked eyes of those who were looking skywards. 'Observation', however, is not necessarily the best word to use because it implies a degree of method and objectivity that did not generally exist in an era that was more likely to look towards mystical and magical explanations for natural phenomena than anything that we would recognize as scientific. It is probably more accurate to say that in pre-scientific days, comets were undoubtedly *seen*, but they were not necessarily *observed*. But the matter goes further than that. The lens through which comets were seen was that of a deeply superstitious era, a time in which people did not necessarily expect a rational explanation for natural phenomena. The world – indeed, the universe – manifested the whims and the moods of the gods and, as such, natural phenomena might portend good or ill, depending on those moods. Correctly interpreting these phenomena could give valuable insights into the gods' temperaments and could help a person or a village to survive and, with luck, to thrive.

Ancient astronomers would have noticed that there are two types of object that always appear in the skies – those that are fixed and never move with respect to each other (the stars), and those that move against the starry background (the planets, the Sun and the Moon). The stars appear in the skies or vanish from view over the course of each night, but also over the course of each year; it is easy to picture them as fixed in the heavens, perhaps attached to some sphere in the far distance. The planets move across this background, but they all move in a predictable manner – their motions are largely confined to a narrow band (the ecliptic), and they are repetitive with time; they all return to the same location relative to the background stars on a regular schedule that can be predicted, even if that schedule plays out over decades. The important part of all of this is that for those who watch the skies, the rising and setting of stars and the motions of planets are regular and predictable.

Now imagine that one night there is something new. It probably shows up outside of the ecliptic, so it is not in the part of the sky occupied by the planets. And since there is no record (oral or written) of it ever having appeared before, it is most likely not a planet. Over the course of time – probably in just a few weeks – it will move against the stars, making it clear that it is not a star. Next it starts to change its appearance – something else that planets and stars never do. It might become fuzzy and grow a tail. Nobody has ever seen anything like this before and nobody knows what this new object is, just that it is in the sky.

As the early skywatchers tried to figure out what this new object was, they probably considered what properties they could observe. The fact that it moved meant that it could not be a star, but its appearance outside of the ecliptic would suggest it was not a planet. The fact that it would look like nothing else in the sky – and that it changed its appearance from week to week – might make one think this object is not of the heavens. So what options are left?

In the third and fourth centuries BCE, some of the ancient Greeks used this chain of logic to conclude that comets were not of the heavens – at least not entirely.[1] Hippocrates (460–370 BCE), for example, hypothesized that a comet's tail formed when the comet attracted moisture from the Earth beneath it, forming a cloud that reflected the sunlight. So while the comet itself might be in the heavens, the comet's tail was an atmospheric phenomenon. Taking this line of thinking a little further, Aristotle (385–322 BCE) speculated that this meant the conditions that gave rise to comets' tails might also give rise to drought and high winds and that the presence of comets could be used to predict the weather. For this reason, Aristotle wrote about comets in his book *Meteorologica* (in which he considered comet tails to be 'dry and warm exhalations' of the atmosphere) and not in his book about the heavens, *De caelo*.

The Romans – primarily the natural philosophers Seneca (4 BCE–65 CE), Pliny the Elder (23–79 CE) and Ptolemy (90–168 CE) – also studied comets. For the most part, Roman thinking was not markedly different from that of the Greeks, although the Romans did make some advances. Seneca, for example, had the temperament of a scientist in spite of living in an era of magical thinking, an era in which the prevailing wisdom made it difficult to conceive of a universe that could hold objects that were neither planets nor stars, nor of the Earth. In spite of the limitations imposed by the time in which he lived, Seneca came up with a remarkably modern view of comets, even suggesting that they were permanent parts of the cosmos (as opposed to temporary apparitions), and he speculated that they move in circular orbits as do the planets. With regards to their irregularities compared to the motions of the planets, Seneca, in his first-century *Natural Questions*, felt

> Nature does not turn out her work according to a single pattern . . . [Nature] does not often display comets; she has assigned them a different place, different periods from the other stars, and motions unlike theirs. She wished to enhance the greatness of her work by these strange visitants whose form is too beautiful to be thought accidental.

A tablet from Babylonia, now in the British Museum, discussing the 164 BCE apparition of Halley's Comet.

Seneca the Younger.
Original engraving
by Lucas Vorsterman
(1595–1675).

Seneca also knew that during a fortuitous solar eclipse, a comet was seen to be close to the Sun, a comet that would normally have been obscured by the Sun's glare. Seneca concluded that comets must reside among the heavenly bodies, but he also understood that his knowledge of comets would be forever limited, writing in *Natural Questions*,

The day will yet come when posterity will be amazed that we remained ignorant of things that will to them seem so plain

... Men will some day be able to demonstrate in what regions comets have their paths, why their course is so far moved from the other stars, what is their size and constitution. Let us be satisfied with what we have discovered, and leave a little truth for our descendants to find out.

Pliny the Elder was more of a naturalist than a scientist. He observed and described the world around him without necessarily trying to explain it as a scientist would. Pliny wrote about virtually all aspects of the natural world, and his writings on comets were fairly voluminous. Much of Pliny's writing consisted of repeating what he had heard or read, and, unfortunately, he tended to be somewhat credulous and uncritical in his approach to these tales – in Pliny's case, since his writing was so well respected, this meant that what amounted to tall tales were given an imprimatur that lasted for centuries, into the Middle Ages.

Pliny wrote, for example, that comets appeared mostly in the north, towards the direction of the Milky Way. He also wrote that comets in the southern skies tended to lack tails, and that not all comets moved across the sky. Pliny gave a great deal of credence to comets' role as portents, noting a number of incidents in which a comet's appearance was followed by disaster or ill fortune. Towards this end, he also gave tips on how to interpret a comet's appearance – the shape and direction of its tail, for example, or its location in the sky, and other factors that could be used to determine what might be coming. Furthermore, he repeated a classification scheme with at least ten categories of comets; as with so much else that Pliny reported, this scheme seems to have originated elsewhere (most likely with Seneca) and was simply echoed by Pliny. According to Pliny, comets could be Pogonias (bearded or with a mane), Acontias (javelin-like), Xiphias (shaped like a dagger), Disceus (amber-coloured and disc-shaped), Pitheus (cask-shaped and smoky in colour), Ceratias (horn-shaped), Lampadias (torch-like), Hippeus (like the mane of a running horse), Argenteus (silver-coloured and hard to look at) and Hircus (goat-like).[2]

Pliny also commented on the nature of comets, apparently borrowing from the writing of Seneca (albeit without crediting Seneca in his book). In his book *Natural History*, Pliny wrote,

> Some persons think that even comets last forever, and that they travel in a special circuit of their own, but are not visible except when the sun leaves them. There are others, however, who think that they spring into existence out of chance moisture and fiery forces, and are then dissolved.

One of the greatest astronomers of the classical era was Ptolemy, whose masterpiece was the *Almagest*. In this volume, Ptolemy aimed to explain the motions of the planets through the sky; the method he developed, although ultimately shown to be wrong, was accurate enough to last for more than a millennium. Not until the seventeenth century was it superseded by a better system. When it came to comets, though, Ptolemy set study back significantly; because comets failed to follow the rules of motion he developed for planets, he felt they must be supernatural in nature. As a result, comets fell into the realm of astrology and not astronomy – for the next fifteen centuries, comets were felt to be omens, portents and signs from another realm.

Seneca's observations and theorizing had been a remarkable advance in thinking about comets, and Ptolemy's mathematical treatment of planetary orbits (although ultimately proven wrong) began to move astronomy towards being a mathematically predictive science. But Ptolemy, in viewing comets as supernatural, was wrong, and it reversed much of the progress made by Seneca. His views, in fact, were the beginning of a regression to Aristotelian thinking, and this way of thought became the norm through the medieval centuries. Aristotle's work did not contradict the Bible or Church teachings; for this reason, the Church accepted Aristotle's views on nature, and understanding of the natural world ground nearly to a halt for over a thousand years. Comets were again thought of as atmospheric phenomena, as conveying messages from an angry and cryptic God.

This is not to say that nothing at all happened during these years – towards the end of the medieval period and a century or so before the Renaissance, there were those who began to question the Aristotelian orthodoxy, who began to look for more rational explanations. As noted by the science historian Jane Jervis in her engaging book *Cometary Theory in Fifteenth-century Europe*,[3] one of those who carefully examined some of the ancient theories of comets was Albertus Magnus (1193–1280) in his book *Meteorology*. The majority of Magnus' writing on comets was devoted to a discussion of Aristotle's theories of comets, largely agreeing with what Aristotle had concluded about their origin and nature. But Magnus also added a number of digressions in which he gave his views on Seneca and a number of unnamed 'moderns' who had also weighed in on comets. Magnus went on to classify comets – interestingly, based on the texture of the vapour in the comets' tails rather than by colour,

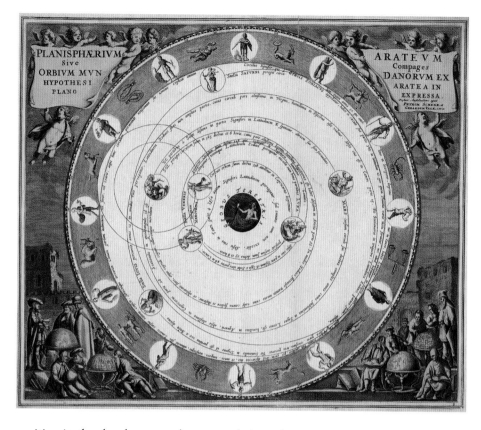

position in the sky, shape or other more obvious characteristics. He closes with an examination of comets' astrological significance, in which he concludes they have little to say about the coming of wars or the deaths of kings 'since vapor no more rises in a land where a pauper lives than where a rich man resides, whether be king or someone else'. Continuing to discuss comets' astrological impact, Magnus decides that they must be caused by the influence of Mars, which then must determine their fate, that since Mars 'is the cause of war and destruction of peoples . . . so a comet is said to signify these . . .'

By the fifteenth century, observers were beginning to make careful observations; these observations were not adequate in and of themselves to develop a scientific explanation of comets and their orbits, but they did lay the groundwork for the

Depiction of planetary positions of 18 March 816 CE (based on a Ptolemaic model).

advances that were to come. The Viennese astronomer Georg Peurbach (1423–1461), for example, wrote a letter and a later treatise (both described in Jervis's book) in which he tries to determine the distance to and size of the great comet of 1456 (now known to be that year's apparition of Halley's Comet). While Peurbach made a number of mathematical errors and used flawed assumptions, he made a legitimate effort to calculate these characteristics as opposed to simply guessing. In addition, Jervis notes that Peurbach seems to have been the first person to try to use parallax to determine the distance to a comet (parallax is the process of observing the shift in an object's apparent position when viewed from different angles), although this attempt was in vain.

Chinese astronomers were also active during these years, but their work was more documentary in nature. They observed the skies and noted what they saw, but they seemed less interested in trying to explain their findings than with predicting eclipses, the positions of planets and trying to determine the significance of celestial events (were they messages of some sort? Did they

Chinese report of the 240 BCE appearance of Halley's Comet.

presage events on Earth? What did they portend?). So while we can look to Chinese records to help confirm that a comet was in the skies – even for records of comets that European skywatchers might have missed – what we cannot do is to look for speculations as to what comets might be.

The ancient Egyptians, although they observed comets, never seem to have developed a theory of where they came from or what they might signify – this is not to say that they had no such theory so much as, if they did, it seems to have been lost through the millennia. This is ironic in that a piece of natural glass found in a brooch of King Tutankhamen's has been traced to a comet that struck the desert in what is now Egypt about 28 million years ago.[4] The artisan who made the king's brooch had no way to know this, of course, nor that a piece of black stone would one day be found in the centre of a veritable sea of natural glass – this stone turns out to be laced with microscopic diamonds, formed by the violent heat and pressure of the comet's impact. This stone, which created the glass that wound up in Tutankhamen's brooch, is also the first found on Earth that has been conclusively shown to be from the nucleus of a comet. Unfortunately, there was no way for the unknown artisan – or for the Egyptian court astronomers for that matter – to know how this glass came to be, and no coherent Egyptian theory of comets has survived.

Of particular interest were some of the early attempts to determine the distance to the comets: if the comets were found to be at any significant distance from the Earth's surface, then they could not possibly be atmospheric phenomena. Although early efforts were unable to provide any resolution to the matter, the Danish astronomer Tycho Brahe (1546–1601) collected enough observational data of the Great Comet of 1577 to determine conclusively that comets not only lay beyond the atmosphere, but beyond the Moon. Tycho's findings were generally accepted, but they still stirred up a debate that lasted for decades after his death.

Tycho did much more than this; as a scientist, his work was impressive, earning him the sobriquet of the last (and arguably the greatest) of the naked-eye astronomers. Tycho was even able to convince the king of Denmark to build him the best observatory available in pre-telescopic times. One of his first major discoveries came at the age of 26, when Tycho noticed a new star in the constellation of Cassiopeia. After careful study, he realized that it was indeed a new star in the sky (we now know it was an exploding star called a supernova) and that it lay beyond the Moon and the planets. This flew in the face of conventional wisdom dating back to the time of Aristotle (384–322 BCE) that held the heavens to be eternal and immutable. Following this discovery, Tycho spent much of the rest of his life making visual observations of stellar positions; using equipment he perfected, Tycho was able to map the heavens with a remarkable degree of precision, better than any of his predecessors or contemporaries.

As alluded to above, Tycho made significant contributions to the understanding of comets, in particular in showing that they were certainly astronomical phenomena. This discovery was made by carefully comparing his observations of the comet to those made by a colleague in Prague. Both observers saw that, while the apparent position of the Moon was slightly different from the two locations, the comet was motionless against the background stars, proving that the comet was more distant than the Moon. On top of that, Tycho observed that the comet's coma was being pushed away from the Sun.

Tycho's life story is about as remarkable as his discoveries; he was, to say the least, a colourful character who had an artificial nose (having lost his in a duel), and a pet beer-drinking moose (or elk – accounts differ). Even Tycho's death was interesting: a guest at a dinner party, he was determined to follow the etiquette of the day and chose to remain at the dinner table rather than rising to urinate; he died less than two weeks later of bladder and kidney problems stemming from this incident. Tycho's were perhaps the last major discoveries about comets in the pre-scientific era – and they paved the way for the science that was to come.

EDMVND. HALLEIVS LL.D.
GEOM. PROF. SAVIL. & R.S .SECRET.

Thomas Murray,
Edmund Halley, 1687,
oil on canvas.

Early scientific era (Kepler, Galileo, Halley, Newton)

Coming from a naked-eye astronomer, Tycho's observations –
and those of his contemporaries and near-contemporaries
– were, at best, quasi-scientific; they collected valuable data,
but they were unable to exploit it fully, a task that was left for
following generations. One of the first to make use of this data
was Brahe's student, Johannes Kepler (1571–1630).[5]

Exactly how Kepler got his hands on Tycho's data is open to
conjecture – Kepler implied that he stole the data after Tycho's

Unknown artist,
Johannes Kepler,
early 17th century.

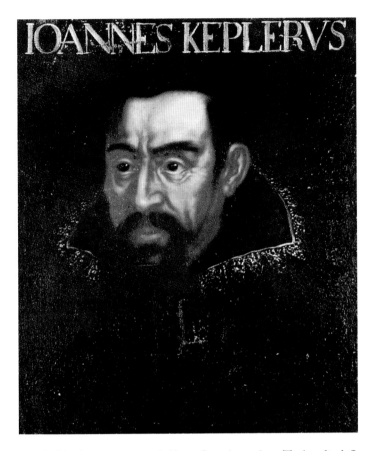

death (Kepler commented, 'I confess that when Tycho died, I quickly took advantage of the absence, or lack of circumspection, of the heirs, by taking the observations under my care, or perhaps usurping them.'), but he was appointed to Tycho's position and might well have had legitimate access to Tycho's data. However he obtained the data, Kepler made good use of it; his analysis showed that planets moved in elliptical orbits as opposed to the perfect circles hypothesized by Aristotle and those who followed him.

Although this seems a minor point, it is actually profound. Consider – if planets were placed in their orbits by God, then those orbits should be perfect circles and not slightly squashed ellipses. And if a planet could move in an ellipse – even one that was very nearly circular – then there was no reason why other objects might not move in much more elliptical paths.

But there was much more to the matter than this: elliptical orbits helped to resolve some of the problems earlier skywatchers had had when trying to predict planetary motions. Perfectly circular orbits should be easy to predict, but there were mysterious variations when astronomers looked at the planets. Not only did some planets (Mars, Jupiter and Saturn) sometimes move backwards across the sky, but even Venus and Mercury showed annoying discrepancies when their actual positions were compared with what was predicted. Ptolemy had suggested

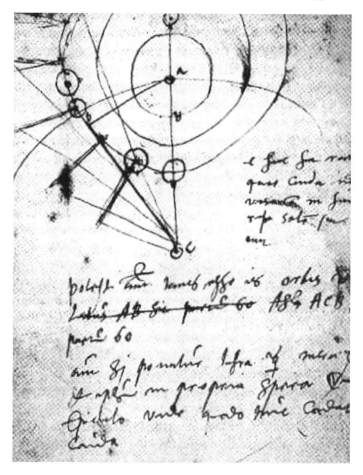

Tycho Brahe carefully observed the Great Comet of 1577, as shown in this sketch from his notebook.

that planets moved in perfectly circular orbits, but that their actual orbits included epicycles – circles upon circles – and these epicycles explained the occasional oddities in the observed motions. The epicycle hypothesis lasted for centuries, and it explained the observed motions fairly nicely, but it was needlessly complex. The only 'universe' in which it was necessary was one in which the planets moved in perfectly circular orbits and the Earth was at the centre of the universe; when Copernicus and, later, Galileo showed that the planets (Earth included) revolved around the Sun, it began to look as though epicycles might not be needed (although there were still annoying orbital discrepancies). When Kepler showed that these orbits were elliptical, all of these discrepancies vanished. In one stroke, Kepler pointed the way to showing not only that the orbits were not circular, but that we could use mathematics alone to predict the motion of the planets. His later observations of comets showed that they, too, tended to move in elliptical orbits, albeit highly elongated ones. Kepler did a wonderful job of figuring out how the planets and comets moved through the skies – what he had not figured out was any sort of central theory that explained these motions in their entirety. That part was left to Newton, but first we need to take a look at Edmund Halley (1656–1742).

Halley, although best known for the comet named for him, had much more diverse interests than one might expect. In addition to his position as Britain's second Astronomer Royal, Halley published research on the weather, climate, atmospheric physics and life expectancy, as well as inventing a diving bell and making contributions that helped make magnetic compasses more reliable. But his most important contribution was in recognizing that comets could return time after time. Until he came to that realization, it was thought that comets travelled in a straight line, passing through the Solar System once, never to return.

Halley's greatest triumph was his recognition that a single comet – what we now know as Comet Halley – had appeared repeatedly throughout history, and he could predict when it would next return. Halley started this line of thought when he plotted the path of the comet of 1683. Using Newton's laws of gravity and motion, Halley realized that, rather than following a straight line through the Solar System, the comet's orbit was closer to a parabola, a path that would carry it into interstellar space and would never return. Halley had actually suggested applying Newton's laws to comets in a letter he sent to Newton in 1687, although it was not until 1695 that he began his own calculations.

Going somewhat further, Halley made a prescient comment in a letter, stating, 'I am more and more confirmed that we have seen that Comett now three times since ye Yeare 1531.' Using astronomical observations and data collected by other astronomers, Halley speculated that the comets of 1531, 1607 and 1682 were three return visits of the same comet; he went on to predict that the comet would reappear in 1758 or 1759. Halley's prediction set off a flurry of activity as astronomers throughout Europe fought to be the first to spot the comet when it returned to the inner Solar System. In spite of the best efforts of the era's leading professional astronomers, the comet was first recovered when a German amateur astronomer, Johann Georg Palitzsch (1723–1788), spotted it on Christmas evening, 1758. Sadly, Halley had died seventeen years earlier.

By this point in time, nearly all of the pieces were in place: Tycho had recognized that comets lay beyond the Moon; Kepler understood that they moved in elliptical orbits; and Halley realized that comets could return repeatedly to the inner Solar System. But what was lacking was a comprehensive theory that explained not just *how* comets moved in the manner observed, but *why*. This was Newton's contribution.

Isaac Newton (1642–1726) was one of the most productive, and one of the most brilliant, scientists in history. Newton's major insights were that all objects in the universe exert a gravitational pull on all others, and that the power of this attraction varied with the inverse square of the distance between those objects – double the distance and the gravitational force drops by a factor of two squared (four). He also realized that the strength of this attraction depended on the mass of the objects – double the mass, and the force will be doubled. These observations were

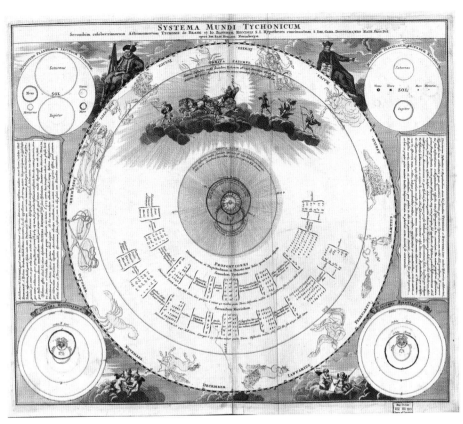

Sketch of Brahe's solar system by the Italian nobleman Peitro della Valle, showing a compromise cosmology with most of the planets orbiting the Sun, which orbits the Earth. Della Valle also shows comets crossing some planetary orbits.

not completely novel; what Newton did was to put numbers to these relationships, along with determining the value of the physical constant that relates the gravitational force to all of these parameters. (For the interested reader, the gravitational constant is 6.674×10^{-11} N(m/kg)^2).

Newton's other major achievement came when he formulated his laws of motion, describing how objects move under various conditions. His first law (an object at rest will stay at rest; an object in motion will stay in motion unless acted on by an outside force) couples with his second law (the change in motion of an object is proportional to the mass of the object and the force applied to it) to show how objects move through space. These laws, combined with his gravitational theory, explained not just how objects on Earth moved, but explained the motions

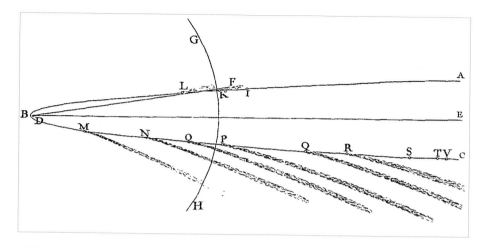

of the distant planets. In fact, Newton's laws are still taught (and used) by astronomers today and by space scientists to navigate our probes through the Solar System. Although he was never able to apply his discoveries outside our Solar System, Newton's laws are genuinely universal. More to the point for this book, Newton's laws of motion and of gravitation made it possible for the first time to understand what it was that astronomers saw in the heavens; they explained why planets (and comets) follow elliptical orbits, why comets move so much more quickly as they approach the Sun, how cometary orbits are influenced by the gravity of giant planets and much more. As mentioned earlier, Brahe, Kepler and Halley showed us *how* comets moved through space; Newton explained *why* and, in so doing, he opened the entire universe to scientific explanation.

Newton plotted the orbit of the great comet of 1680 in his masterwork on physics, the *Principia*, published in 1687.

This is something we take for granted today – that scientists on Earth can understand what is happening on the other side of the universe – but there is no reason why this should be. Why, after all, should the same laws governing an apple falling on Earth be useful in describing the motion of a comet around the Sun, the motion of a star around a black hole or the motion of a galaxy through space? In Newton's day it was perfectly reasonable to believe that the laws of physics that we observed on Earth might not apply to the other planets, let alone to the distant stars. After all, things varied from place to place on Earth

– different continents had different plants and animals, different rocks, different geographies and even different stars showing in the skies. It was plausible to think that the laws of physics might vary from place to place on Earth. When scientists learned that the entire Earth followed the same laws of science, it was still only speculation to think that these laws extended throughout the Solar System; it took a huge leap of imagination to consider that not only the Solar System, but the rest of the universe might be understandable and explainable by laws that humans could think up. As recently as a century ago, Albert Einstein (1879–1955) commented, 'The most incomprehensible thing about the universe is that it is comprehensible.' Kepler and Halley showed us that the universe – or at least our part of the universe – is predictable; Newton showed us that we can apply these same predictions to the rest of Creation. But in spite of his success in predicting comets' motions, Newton was unable to learn what comets were made of and where they came from. That had to wait for nearly two centuries more.

Current era (composition and structure, origins, spacecraft visitations)

In 1842 August Comte (1798–1857), a French philosopher and arguably the first philosopher of science, wrote about the difficulty of ever knowing what the planets are made of, stating,

> Of all objects, the planets are those which appear to us under the least varied aspect. We see how we may determine their forms, their distances, their bulk, and their motions, but we can never know anything of their chemical or mineralogical structure . . .[6]

Comte felt that, barring the ability to visit the planets physically to collect samples, there simply was no way for us to know what they were made of. And since the planets were apparently out of reach, any accurate information of their composition was seemingly impossible to obtain. Comte would never have guessed that, within a handful of decades, he would be proven wrong and that, within a century, scientists would have a fairly detailed understanding of the composition of planets, comets and even the stars.

Today we take for granted our ability to sample remote objects directly – we have landed men on the Moon; landed probes on Mars, Venus and Saturn's moon Titan; and recently even landed on Comet Churyumov–Gerasimenko. Other space-craft have returned samples of the solar wind and of the dust that fills interplanetary space, not to mention the flyby missions that have sent craft past every planet in the Solar System. But, while these missions were inconceivable to Comte, scientists already had a fairly good idea of the overall composition of these objects decades before the first rocket left Earth. The process used to gather this information is called spectroscopy.

The early years of the twentieth century were when scientists teased out the structure of atoms and discovered some of the basic precepts of quantum mechanics; with this knowledge, they began to understand the composition of the stars, planets and other bodies in the universe – comets included. Pioneers such as J. J. Thompson (1856–1940), Ernest Rutherford (1871–1937) and Niels Bohr (1885–1962) worked out the structure of atoms, while Bohr, Max Planck (1858–1947), Albert Einstein (1879–1955) and others who followed learned how an understanding of these properties can help us to identify chemical elements and molecules at a great distance.

Every atom is surrounded by electrons, and those electrons are arranged in shells that are at distinct distances from the atom. Just as it takes energy to move from one stair to the next higher stair, so too does it take energy to move an electron from one orbital to the next. And just as the distance (and the energy) to go from one step to the next is the same, so, too, is the energy required to jump from one orbital to the next. The energy needed to jump from one level (that is, the energy level in an electron shell) to another is unique for each pair of orbitals and for each element; if we can find a way to measure this energy difference we will know exactly what element we have found.

Luckily this is possible because it turns out that we can see these jumps – when an electron drops from a higher to a lower energy level, it emits a very specific frequency of light (this is where the colour of the light in a neon lamp or the Aurora Borealis, the Northern Lights, comes from). Similarly, absorbing a photon with a very specific wavelength will lift an electron into a higher energy level. So if we can measure the exact wavelength, or colour, of the light emitted – or absorbed – by an object, we can find out what elements it contains. And the way to do that is through a technique called spectroscopy.

As Newton showed, passing light through a prism would spread it out, showing all of the colours that comprise white light. What he was unable to see was the incredible amount of detail contained in a spectrum; he lacked the instruments to extract all the information that was there. In particular, what he was unable to see was a network of dark lines lacing the spectrum – each line represents the absorption of very specific wavelengths of

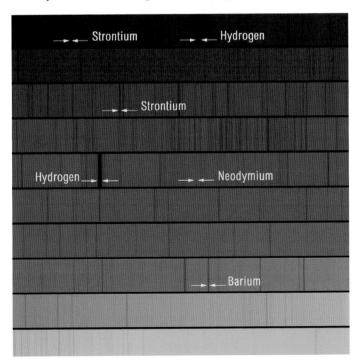

Example of the blue wavelengths of a spectrum showing dark absorption lines.

An article from
9 February 1910,
Ogden (Utah)
Standard predicting
dire effects from
the coming passage
of Earth through the
tail of Halley's Comet.

COMING END OF WORLD

Halley's Comet May Snuff Out Life on the Earth

Berkeley, Cal., Feb. 9.—"If the astronomers are right in their estimations of the amount of cyanogen gas in the tail of Halley's comet, and if that body's vapors do envelop the earth, we may have a chance to feel the sensations of the bugs and insects which are killed by the use of this deadly gas as an exterminator."

light by material between the source of the light and the Earth. By looking at the spectrum of sunlight passing through a comet's tail, scientists can tell what elements are present; by knowing which elements are present and in what concentrations (the strength of an absorption line tells us about the amount of the element in question), scientists can start to make guesses about the molecules that the object contains.

The first person to notice lines in a spectrum was the German optician Joseph von Fraunhofer (1787–1826); Fraunhofer fabricated the purest glass prisms of the day, and, when he passed sunlight through his prisms, he noticed an array of thin, dark lines in what he expected to be a continuous gradient of colours – 574 lines in all. Decades later, the British-American astronomer Cecilia Helena Payne-Gaposchkin (1900–1979) showed in her doctoral dissertation of 1925 that these Fraunhofer lines in the solar spectrum could be correlated with specific chemical elements; one of her advisers, the American astronomer Henry Norris Russell (1877–1957), extended her work and applied these results to other stars and even galaxies. What these scientists discovered was that each chemical element had a specific 'fingerprint' of spectral lines that could be used for identification.

So say that scientists peering at the light passing through a comet's tail find that it absorbs wavelengths of light associated with hydrogen and oxygen – it seems reasonable to assume that water vapour (water is composed of two atoms of hydrogen and one of oxygen) is present in the comet's tail. Since the tail comes from the comet itself, it further seems reasonable to assume that the comet is made at least partly of water ice. If carbon and hydrogen are found, then methane is also likely present. But these molecules have their own spectra, and astronomers can also look directly for the methane spectral line. Some molecules have spectral lines in visible light, but they are more often found at longer (lower-energy) wavelengths in the infrared or radio wavelengths. Examining the spectrum of Comet Halley during its appearance in 1910 is what told astronomers that cyanogen was present, a discovery that caused some degree of panic when it was announced.

Once spectral analysis was developed, astronomers could learn a great deal about what comets were made of – the volatile parts, in any event – and they were able to link some meteor showers with parent comets; recovering some of these meteorites gave insights into the composition of the rocky parts of the comets. Most of the big picture was in place, but they still lacked information on how comets were constructed. In the absence of any close-up images, the astronomers simply did not know how much of the comets were comprised of ice and rock (or dust), and how they were put together. These last bits started to come together towards the middle of the twentieth century.

In 1949 the American astronomer Fred Whipple added yet another piece to the puzzle when he suggested that comets were made mostly of ice, not rock and dust. The fact that ice was present was obvious – there was no other way that comets could sprout such magnificent tails. But most astronomers were convinced that ice alone would evaporate completely during a passage close to the Sun; the fact that sun-grazing comets were seen to survive their passage seemed to indicate that comets must be mostly rock. But if this was the case, there would not be enough ice to sustain a comet such as Halley's during its repeated visitations. Whipple was able to resolve this dilemma by calculating not only how much ice would evaporate during each passage through the inner Solar System, but by realizing that the majority of the comet would remain cold enough to avoid vaporization – the same phenomenon that lets one make Mexican deep-fried ice cream or the dessert dish baked Alaska. Whipple also proposed a model describing how comets' surfaces change with time, becoming dustier as ice evaporates, with jets of gas blasting dust and other solid materials into space to become the tail. In the words of astronomer Michael Belton, commenting on Whipple's landmark paper of 1949, Whipple's model

EARTH GOES UNHARMED THROUGH COMET

——

Switched by Tail of Celestial Vagrant With the Gentlest Sort of Caress

——

EIGHT HOURS IN THE TENUOUS COSMIC VAPOR

——

Fears of the Superstitious, Who Looked for Dreadful Catastrophes, Prove Baseless

——

HUGE SPOTS FRECKLE THE FACE OF THE SUN

——

Headline from the *San Francisco Call*, 19 May 1910, after Earth passed through the tail of Halley's Comet in 1910.

signaled a fundamental turning point in our understanding of the chemical and physical nature of comets and, thereby, our appreciation of their value in probing the conditions

Example of a 'dirty snowball' from the classroom of Professor Richard Pogge at Ohio State University.

and processes that occurred in the primitive history of the solar system.[7]

The last piece of the big picture was provided at about the same time, when both the Dutch astronomer Jan Oort (1900–1992) and Estonian astronomer Ernst Öpik (1893–1985) proposed the existence of a spherical shell of icy bodies surrounding the Solar System at a great distance from the Sun – what we now know as the Oort Cloud. Oort and Öpik suggested that these objects, the detritus left over when the planets and moons formed, were cast out of the inner Solar System by Jupiter's gravity and that of the other giant planets; scattered from their birthplace in the ecliptic to form a distant cloud in the outermost reaches of the Solar System. Drifting here for hundreds of millions – indeed, billions – of years, these icy objects form a reservoir of material, becoming comets when they approach the Sun.[8]

With these two pieces in place, astronomers had learned nearly all they could about comets without being able to obtain actual samples or to image one from a short distance – for these, they would have to wait until the passage of Comet Halley in 1986 and the later spacecraft. What is remarkable is how well the earlier work holds up, given that Whipple, Oort and Öpik never had the chance to see a comet up close.

3 Visualizing Comets

Comets have captured the attention of artists for millennia – we have found Paleolithic images of comets etched and painted on the walls of caves, and they have appeared in works of art across eras until the present day: the great comet of 1066 appears in the Bayeux Tapestry; Giotto di Bondone (1267–1337) painted comets into frescoes in the thirteenth and fourteenth centuries; and comets appeared in paintings of the Renaissance and later centuries. Whether appearing as portents, as messengers, for historical accuracy or simply because of their beauty, they are unmistakable wherever they occur.

According to the Oxford Dictionary, art is

> The expression or application of human creative skill and imagination, typically in a visual form such as painting or sculpture, producing works to be appreciated primarily for their beauty or emotional power.

Typically this is interpreted to include paintings, sketches, sculpture and other products of the creative process, but there is more to human creativity – and to art – than this. Does it matter if a sketch is made to show an artist's interpretation of a comet's visitation as opposed to reflecting what a scientist sees through the eyepiece of his or her telescope? For that matter, does it matter if a beautiful photo is taken for the purpose of science instead of simply to be beautiful? The fact is that scientific research is as much an aspect of human creativity as is art;

both are ways of trying to interpret the world and the universe around us. For that reason, this chapter will not only discuss the paintings and artistic sketches that depict comets, but will include sketches, photographs and even digital renderings of comets made in the course of research.

Antiquity

If you ask a group of disparate scientists which science is the oldest, many of them will make a convincing case for their own discipline. But perhaps the most compelling cases are made by astronomers and geologists – those who look upwards for a living, and those who look down. A geologist will say that the earliest humans had to study the rocks to make the tools and implements that were so important for their survival, and we have ample evidence of their interest in the form of so many of these tools. Astronomers will say that the earliest humans must have looked upwards from time to time and wondered about the lights in the skies above them; we also have evidence of this in the form of petroglyphs left by Stone Age cultures in various parts of the world that show the skies. Interestingly, some of these petroglyphs include images of what seem almost certainly to represent comets.[1]

Surprisingly, a fairly common image in prehistoric art is a swastika. Both the American astronomer Carl Sagan (1934–1996) and the Portuguese anthropologist and geologist Fernando Coimbra made a compelling case that this could very well represent a comet, with the hooked arms representing the tails (ion and dust) and possibly jets emanating from a comet's nucleus; when one compares some drawings and photos of a comet with swastika images, this seems a plausible interpretation. There has even been speculation that this swastika motif must have arisen during a particularly close appearance of a major comet – close enough for the naked eye to see these details.[2]

To start, we need to remember that, before its appropriation by the Nazi Party in the 1930s, the swastika did not have its current association with evil; in fact, even today it is a positive

symbol in India (specifically, among the Jains), and it shows up in imagery from all over the ancient world, from Japan to India, to Tibet, to the buried city of Troy and to Native American cultures. There is more to it than that, though; the swastika is not a simple shape that can be expected to arise spontaneously in the imagination – especially not in so many disparate locations. In the words of Carl Sagan and his wife Ann Druyan,

Petroglyph along the Peñasco Blanco Trail in the Chaco Canyon.

> All of these difficulties seem to be resolved if there was once a bright swastika rotating in the skies of Earth, witnessed by people all over the world . . . we need only examine the sketches and photographs of the spraying fountains in a cometary nucleus, recorded by generations of astronomers, to realize that there is here the potential for generating such a prodigy.[3]

Sagan and Coimbra make a compelling case that, for a somewhat complex image to be so widespread, it must have been inspired by a celestial phenomenon, visible to the whole world;

this phenomenon might well have been a comet that passed so close to the Earth that its nucleus and its jets of gas and dust were clearly visible from Earth. Jets alone were likely not enough to form the hooked cross shape of the swastika but, with the comet's rotation, they would have been bent into the familiar shape. Sagan and Druyan even go as far as to suggest that this must have taken place sometime during the second millennium BCE, based on the observation that swastikas are absent from global art before that time.

But petroglyphs contain more than swastikas – there are a number of images that are even more obviously inspired by comets. There are some general themes that come up when ancient civilizations (the Greeks and Egyptians, for example) develop imagery and mythology associated with comets. In particular, the long and narrow shape of a comet and its tail were echoed in the long and narrow shapes of snakes and rivers in some ancient artwork. But in the photos of petroglyphs shown here, we can see that, even without becoming fanciful, it was still fairly easy for ancient cultures to communicate across the ages that an impressive comet had appeared in their skies.

Field sketch of petroglyph showing the outline of the comet more clearly.

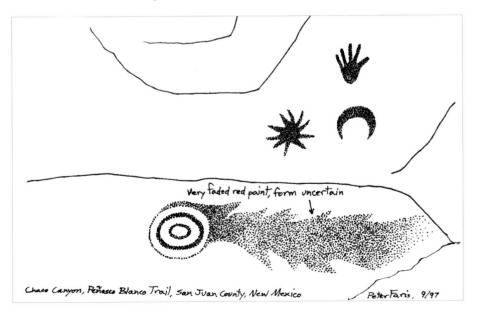

Very faded red paint, form uncertain

Chaco Canyon, Peñasco Blanco Trail, San Juan County, New Mexico

Peter Faris, 9/97

Interestingly, Coimbra points out that comets seem to have been more common in the past than they are today and suggests that this could simply reflect the fact that, with darker skies (no streetlights) and fewer distractions (no buildings), comets were simply easier to notice. And thinking about it, for most of humanity's time on Earth, there have been few – if any – visual distractions in the night sky; the ancient Greeks and other civilizations used oil lamps to light the streets of their cities, but these gave off only limited amounts of light and were restricted to the cities. Once outside the cities, and even in unlit streets, the skies were undiluted darkness. Today, by comparison, the ubiquitous streetlights can drown out faint objects up to 80 km (50 miles) from a city, and many highways and motorways are well lit even further from town. Those looking upwards before our current well-lit era could see everything that appeared in the skies above; the inhabitants of large cities today can see little more than the Moon at night – and sometimes not even that if there are tall buildings around.

The astronomers of ancient China kept meticulous records of what they saw in the sky, including drawing the comets that

A section of the 2nd-century BCE Chinese silk manuscripts showing a variety of comet shapes, including a swastika. From the Mawangdui tombs, Chansha, Hunan Province, China.

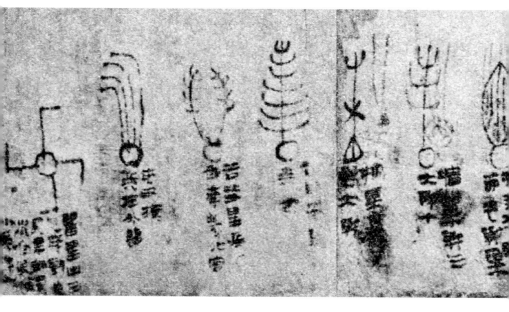

Photograph of the nucleus of Comet Hyakutake with a shape suggestive of a swastika. Image taken by the ESO's New Technology Telescope during its 1996 appearance.

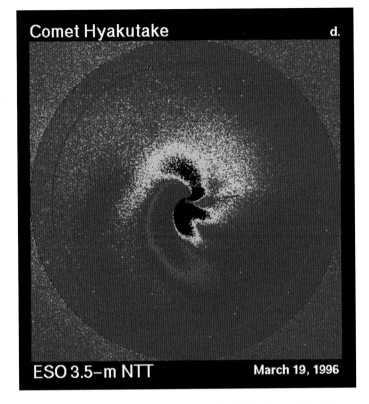

appeared. They left a record of wonderfully drawn depictions on silk, and these drawings are now a good catalogue of the various forms that comets can take. Looking at their drawings, we can see their interpretation of cometary shapes – the multiple tails, jets of gas and striations in the tail (we see this, too, in the photograph of 2007's Comet McNaught shown in Chapter Six).

Jumping forward in time to the Classical era, there were not many representations of comets in art – virtually none in Greek art and very little in art from the Roman era. Romans, in the years following Julius Caesar's death, minted a number of coins with a comet on the reverse to commemorate the comet that blazed in the skies after his death. And within the Roman Empire, adherents of the cult of Mithras carved a gemstone with multiple depictions of comets. But works of art that feature or include comets are rare during Classical times.

Sketch of an amulet from the cult of Mithras. The dagger-shaped objects in the sky have been interpreted as possibly representing comets. This sketch originally appeared in Franz Cumont's monumental treatise *Monuments figurés sur les mystères de Mithra.*

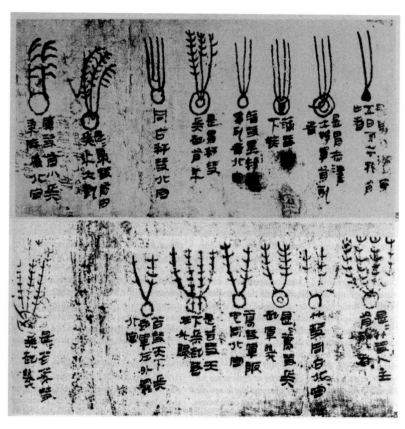

Medieval artwork

Comets start to show up much more frequently during medieval times, typically with religious or mystical connotations. The classic example of this is the Bayeux Tapestry, which shows Halley's Comet in the skies at the time of the Battle of Hastings in 1066. In this tapestry, the comet is clearly seen by the population, who seem to fear what it might portend. A messenger brings news of the comet to King Harold, who has just succeeded the dead King Edward. Woven into the tapestry to the left of the comet is the Latin phrase *isti mirant stella* (these men wonder at the star), suggesting the sense of wonder evoked by the apparition. In this case, the portrayal of the comet is a portent of the events to come – the Norman Invasion and all that was to follow.

This tapping into the religious vein is evidenced in Giotto di Bondone's (*c.* 1267–1337) fresco *The Adoration of the Magi*, showing a comet blazing in the sky as the Three Wise Men visit the infant Jesus at the time of his birth. In this case, as with the comet that was said to signify Caesar's ascension into godhood, the comet signifies the intervention of divinity in the scene. Interestingly, although Halley's Comet made an appearance around 12 CE, there is no indication that it could have appeared as the star in the east that the Bible says guided the Magi to Bethlehem (which is not to say that the star could not have been a comet, just that it was not Halley's Comet). But having said that, the comet in this painting might well have been inspired by Halley's Comet, albeit the appearance that it made in 1305, when di Bondone was in his late thirties. This same theme appears again in the painting *Christ in Gethsemane*, painted by three Dutch brothers, Herman, Paul and Johan Limbourg (1385–1416). In this painting, at least three comets appear in the sky above Christ's head as he waits in the Garden of Gethsemane on the night of his betrayal; the comets here, again, appear to connote the presence of divinity at this important event, possibly even to indicate God's blessing.

Throughout the medieval period, comets were almost invariably depicted as mystical, divine or supernatural, but they were

A section of the Chinese Mawangdui silk texts from the 2nd or 3rd century BCE depicting various configurations of comets and their tails.

not always shown as foreshadowing the future (as was the case in the Bayeux Tapestry); an illustration in a sixteenth-century chronicle of Swiss history, the Luzerner Schilling, depicts a comet in the sky that is raining blood – the illustration suggests that the comet was involved in (possibly the cause of) the Black Plague during one of the Plague's visitations to Europe. This painting not only shows the rain of blood, but it shows people on the Earth below becoming sick, as well as a two-headed cow, presumably caused by the comet. But even here the comet is shown as exerting a distinctly unnatural effect on the Earth

Giotto di Bondone, *Adoration of the Magi*, 1301, fresco.

– raining the death and destruction of the Black Plague on the ground below. The bottom line is that in this era, people viewed comets as almost entirely supernatural manifestations.[4]

The Renaissance and the Enlightenment

Renaissance depictions of comets began to be as much documentary as they were devotional – among them are included the first sketches of comets as seen through the lens of telescopes. People began to see comets as being a part of the natural world rather than as messengers from the supernatural one. The Renaissance also saw the start of historical analysis – people began looking at how phenomena were understood in the past.

Take, for example, the painting of the Great Comet of 1680 by the Dutch artist Lieve Verschuier (1627–1686). In this painting, the comet's tail arches dramatically towards the zenith at dusk – the tail's length and brilliance suggest that this was a sun-grazing comet – while people in the picture's foreground are making measurements of the comet using cross-staffs, early instruments used to measure angles of objects in the sky. In this painting,

A section of the Bayeux Tapestry showing people pointing at Halley's Comet.

Lieve Verschuier, *The Great Comet of 1680 Over Rotterdam*, c. 1680.

which dates from the early years of the Enlightenment, the comet is the painting's focus, and it has no purpose other than to be seen and studied. This comet is not a messenger from God; it is not a portent of things – good or bad – to come. It is an object to be studied by men who are trying to learn more about it and, by extension, about the universe as a whole. Some may be gazing at it in wonder, but not necessarily in awe; they admire its beauty but they do not fear its meaning because it is an object for examination. It might carry information, but it carries no message. A similar piece, a woodcut of the great comet of 1577, is alike in nature to Verschuier's painting in that it shows townspeople watching a comet (although not using instruments to make observations).

The Limbourg brothers, *Christ in Gethsemane*, c. 1415, from *Très riches heures du Duc de Berry*.

Another example of Enlightenment art as part of the scientific study of comets is the sketch showing the progression of the same comets along its orbit as it passed through the inner Solar

75

System. Interestingly, this sketch was made by Isaac Newton for his masterwork, the *Philsophiae naturalis principia mathematica* (usually shortened to, simply, the *Principia*). In his book, as suggested by the title, Newton laid out the mathematical principles underlying the physics he had developed to explain the motion of objects on Earth and elsewhere in the universe. In the time before photography and without much in the way of drawing tools, books such as Newton's included a host of sketches, which served both documentary and illustrative purposes; the sketch of the orbit of the comet of 1680 served both purposes in that it documented the path of the comet through the sky while, at the same time, illustrating the principles Newton was elucidating in his book.

These examples make it clear that comet-related artwork had changed greatly since medieval times. That being said, these centuries still included a number of the more fanciful depictions of comets, including a woodcut illustrating a number of different ways in which comets manifest themselves in the heavens. This illustration, which appeared in Alain Manesson Mallet's book *Description de l'universe* of 1683, shows some comets that are drawn in a somewhat realistic manner side by side with two that are drawn as celestial daggers. Another work of this era shows comets crashing through the zodiac; these depictions do not show comets as mystical agents of God (or the gods), but neither do they depict comets as they actually appeared in the sky – they occupy some middle ground between the real and the mystical. That could, in fact, be a nice metaphor for the Renaissance, a period in time standing between the mysticism of earlier ages and the scientific rationality of later times, a period in which not just scientists, but people in general began to believe that the universe might be explicable and understandable by humans.

Jiri Daschitsky, *The Great Comet of 1577 Over Prague*, date unknown, engraving.

fig 52

Modern era

As we move into the modern era (and especially towards the present day), we notice a number of different threads in art associated with comets. In particular, with the advent of photography and, more recently, computers and digital technology, our ability to capture images of comets – for both scientific and artistic purposes – has undergone a sea change. But more on that in a moment; we should begin with an earlier part of the modern era.

As science began to learn more and more about comets (not to mention the stars and planets), astronomers began to look for them more and more and to see them further and further away from the Sun; as they learned to track comets to the outer Solar System, they realized that comets only have tails when close to the Sun. Further out, they might show a fuzzy coma but not much more. However, it turns out that there are a number of fuzzy things in the sky; today we know them to be nebulae (clouds of gas and dust), star clusters and galaxies. Some comet-hunters of the eighteenth and nineteenth centuries began making lists of fuzzy objects that were *not* comets to avoid wasting time observing them – among the most famous list is the one developed by the French astronomer Charles Messier (1730–1817). Astronomers today continue to refer to many celestial bodies by their Messier designations: the Andromeda Galaxy is M31; the Great Orion Nebula is M42; M1 is the Crab Nebula and so forth. Messier was not the only person compiling such catalogues; one of the more important such catalogues of the mid-nineteenth century was developed by the Americans Elijah Burritt and F. J. Huntington. Not only did their atlas of the skies include maps of the stars and drawings of the constellations, but it included charts to help distinguish between comets and fuzzy objects that might be mistaken for them.

In the early years of the twentieth century, photography began revolutionizing astronomy, and, during the return of Halley's Comet in 1910, astronomers took the very first photos of a comet through a telescope. These photos added

Types of comets, some realistic and some fanciful, depicted in an illustration from Alain Manesson Mallet's *Description de l'universe* (1683).

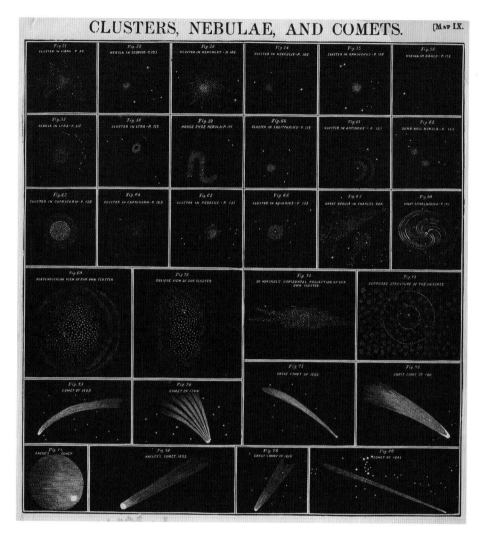

Burritt-Huntington Chart of Clusters, Nebulae and Comets from *Burritt's Atlas to Illustrate the Geography of the Heavens* (1856).

tremendously to the study of the comets: using long exposure times made it possible to capture details invisible to the human eye, and photographic plates enabled scientists to conduct objective observations unaffected by bias, poor drawing skills or lack of attention to detail. But these photos are more than scientific observations; many are beautiful in their own right, and, as photography grew more sophisticated, the photos grew both more revealing of the nature of comets and more striking.

Cartoon from the *Tacoma Times* by Johnny Gruelle, after the passage of Earth through the tail of Halley's Comet on 18 May 1910.

NEVER TOUCHED ME!

Comet Halley's apparition in 1910 also inspired a number of editorial cartoons commenting on the upcoming appearance and potential risk (during its approach to the Sun), and on the Earth's perceived near-miss after passing through the comet's tail. Although the first cartoons featuring comets were editorial in nature, it did not take long for these cartoons to become humorous, poignant or insightful, or even for the comets to be used metaphorically to make a completely non-comet-related political point. The bottom line is that, today, comets can be used symbolically, for comic effect, or to make metaphorical points, and the general public is so familiar with comets that these references are easily understood.

Most comet-related art today seems to fall into one of a few categories: art in which the comet is incidental, added (often along with ringed planets) for dramatic effect; art in which the comet is the main object of an objective piece of art (that is, artwork that is not spiritual or mystical in nature); and art in which the comet is stylized or cartoonish. While there remain depictions that are frankly religious in nature, these are no longer as prevalent as they once were – to a large extent, the religious aspect of comet-related art has been replaced with a more nebulous spiritualism and mysticism that lacks any specific religious symbolism. In recent years, we have seen comets used as devices in popular art – stylized depictions of comets used in cartoons, for children's shows and so forth.

The sun-grazing comet Bradfield Lasco. The photo was taken by NASA's Solar and Heliospheric Observatory on 18 April 2004.

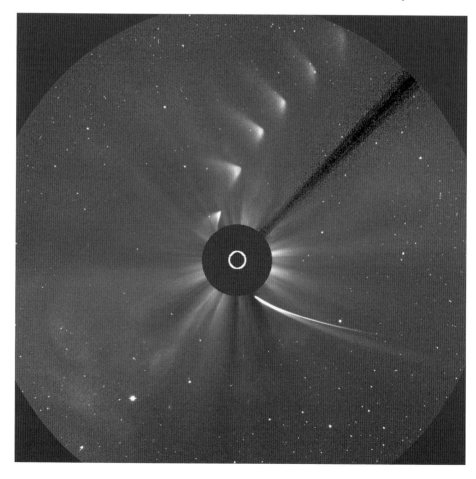

Three views of Comet
c2011L4 taken by
amateur astronomer
Gregg Ruppel.

In recent decades, there has been an explosion in the use of more technologically advanced imaging and image-processing. This makes it possible to obtain images of comets that are far more detailed than anything that was ever possible in the past, as well as obtaining photos from vantage points that were never before possible. There are now beautiful and dramatic photos of comets taken from space (both from the Space Shuttle and from the International Space Station), as well as close-up photos of comets taken by spacecraft sent to rendezvous with them. But our new capabilities go far beyond the ability to take pictures that could never have been taken before. Computers have also enabled the processing of these images to tease out details that were never before visible – subtle structures in the coma and tail, identifying specific elements by the light they emit, and more. The SOHO (Solar and Heliospheric Observer) satellite has even allowed scientists to image comets that could never have been seen before – smaller comets passing too close to the Sun to be seen except by this highly specialized satellite observatory that is specially designed to mask the light from the Sun and make objects close to the Sun visible.

Although most of the false colour images are processed to reveal details that are of interest to scientists, they make striking photos, speaking artistically. And since personal computers are growing increasingly powerful, virtually anyone working at home has the ability to perform image-processing that was previously the domain of scientists at universities and government research centres.

In fact, one can make an argument that these images are art, whether they are intended to be or not, just as photographs from scanning electron microscopes have been recognized for their artistic merit. One of the roles of an artist is to help us to see things in a new way. Artists can do this by altering colours or shapes, by juxtaposing elements not normally found together, by showing details not normally seen in an object and more in order to, in many cases, reveal the hidden essence of their subject or hidden truths about the world. All of these techniques are also used by astronomers and other scientists to help to understand their subject better. But science and art have even deeper similarities than this – many aspects of the creative process are virtually the same, whether what is being created is a painting or a theorem. In spite of the popular distinction between 'left brain' (logical) and 'right brain' (creative), the fact is that creativity in science and art – normally considered 'left brain' and 'right brain' activities, respectively – have much in common. To a large extent, the act of creation is the same, regardless of what is being created: scientists talk about the beauty of theorems, proofs and even equations with the same passion that artists of all sorts speak of the beauty of colours on canvas, or musical progressions, or of the written word. And both camps appreciate the beauty of the natural world: images from electron microscopy are routinely sold as art, as are telescopic photos and even shapes or objects drawn from mathematical equations (fractal art, for example, is derived from a mathematical equation developed by the French mathematician Benoit Mandelbrot). Science can be – and is – creative, and as we see here, the fruits of scientific creativity can be every bit as beautiful as the fruits of artistic creativity.

Returning to the topic of the portrayal of comets in art, by now it should be clear that comets can be portrayed in any number of ways and for any number of purposes. Whether the purpose of a picture (painted, sketched, photographed or computer-generated) is to show comets delivering the judgement of heaven, to convey important scientific truths or simply to look pretty, artists have used comets in their work for millennia and will likely continue to do so for as long as humans exist. They are simply too beautiful, too striking and too important to ignore.

4 Comets and Religion

When Julius Caesar was assassinated in 44 BCE, the Roman Empire was on the brink of civil war; within a decade, the empire was relatively stable and under the control of Caesar's adopted son, Octavian (later Augustus Caesar). Part of the reason for this was not only that Caesar was declared a deity by the senate of Rome, but the appearance of a great comet in the skies confirmed his deification in the eyes of the citizens of Rome. Appearing in the heavens, comets have inspired religious awe as well as religious dread, and these feelings are not limited to antiquity. As recently as 1997, the Heaven's Gate religious cult committed mass suicide, driven in part by their conviction that the great comet of that year (Hale–Bopp) heralded their chance to move on to a higher level of being. Finally, there is a view of comets that is remarkably consistent – whether viewed through the lens of animism, astrology, polytheism or monotheism, comets are most often seen as carrying messages from the heavens to the Earth.

In this chapter, we will take a look at how comets have been viewed by the world's religions over the ages – how they have inspired hopes and fears, and how they have fitted into (or not) the prophecies of the day.

Astrology

Ancient zodiac wheel mosaic from a 6th-century synagogue in Beit Alpha, Israel.

It was not long ago – a few centuries – that astrology was considered a science and astrologers were considered to be among the most learned of learned men. For millennia astrologers had places in royal courts and were consulted by the rich, famous and powerful; they were asked to weigh in on everything from personal problems to affairs of state. Astrologers did much of what we picture scientists, especially astronomers, doing: they carried out complex calculations; they observed the heavens; they made predictions that they checked against the real world; and they developed elaborate theories to try to explain why their calculations and predictions did or did not work. Today we would say that astrology is a pseudoscience; it has all the trappings of science, but it relies on premises that cannot be checked – the premise, for example, that heavenly bodies can actually affect events on Earth. Astrology is, today, not considered a science, but is it a religion? It certainly shares some characteristics with religions – namely that its fundamental premise must be taken on faith because it cannot be rigorously tested.

The central premise of astrology is that the heavenly bodies influence us and our destinies; by understanding the positions of the planets and other heavenly bodies at the time we came into the world and at specific times after that, astrologers feel they can

The Italian cartographer Geminiano Montanari's sketch showing the track of the great comet of 1664 as it passed through constellations of the zodiac.

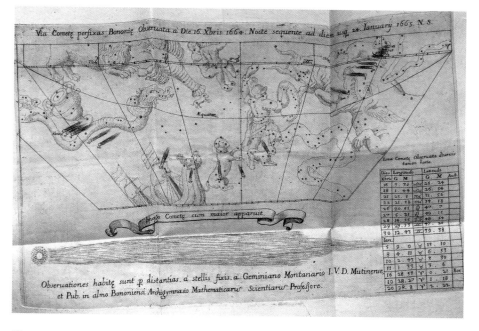

predict what will happen to us and can recommend the correct course of action at critical junctures in our lives. If heavenly objects are related to events on Earth, then the appearance of a rare and spectacular visitor that can appear far from the ecliptic is bound to be an important omen. The role of the astrologer was to figure out exactly what this omen meant.

One of the most candid comments about comets was written by the astrologer Jonathan Flanery in an Internet article from 2005 titled 'Unexpected Visitors: The Theory of the Influence of Comets'.[1] In it, Flanery states:

> Except for a few periodic comets and regular meteor showers, these apparitions are unexpected and we astrologers are usually at a loss as to what they signify except perhaps with hindsight.

In other words, through most of history, when a comet appears unexpectedly about all that the astrologers could say was that something important was about to happen, but not necessarily what that would be until after it happened. Comets could be used to *retrodict* (that is, to recognize the significance of events that had happened when the comet was in the sky in the past), but not necessarily to *predict*.

In an email discussion during the writing of this chapter, Flanery offered additional information on the role of comets in astrology:

> Most astrologers don't really study [comets] because they are 'cosmic wild cards', their appearance and visibility are unpredictable. In some cases they will suddenly light up as did the comet of October 2007 and sometimes they will be duds and, due to planetary perturbations, sometimes they will not be 'on schedule' . . . From an astronomical perspective, these bodies are too small to exert much force in the way of gravity, they are irregular, and not able to reflect much light. Astrologically they are irrelevant unless they become visible . . . These factors are a large part of why

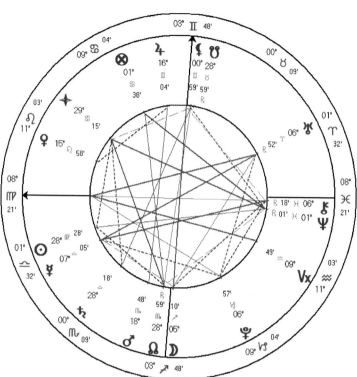

Astrological chart for the time that Comet ISON was discovered.

they are considered to 'carry messages' rather than act as governors as the planets do.

Second is their astrological nature. Their message is delivered when they reach perihelion [the point of closest approach to the Sun] and activated when either the Sun or Mars subsequently crosses the degree of perihelion ... Astrologically, their influence is generally considered negative but, some benefit and some suffer, certainly there is a change in patterns ...

Flanery goes on to give some examples of comet arrivals, noting the appearance of Halley's Comet at the time of the Norman Invasion, as well as its appearance at the time of the Chernobyl nuclear reactor accident in 1986. He also notes that Comet Kohoutek enjoyed its best visibility in the southern hemisphere in a position that he felt was linked to both Saudi Arabia and Chile – close to the time of a coup in Chile and the formation of OPEC, of which Saudi Arabia is a prominent member.

For most of history, comets were considered to bear nothing but bad news. The American science fiction author Isaac Asimov (1920–1992) brought this up in his foreword to Roberta Olsen's book *Fire and Ice: A History of Comets in Art* (1985). Asimov notes that:

> The most obvious unpredictabilities in the sky are represented by the comets . . . A comet with its long sweeping tail looks like the head of a woman with long streaming hair (a traditional sign of mourning), or like a sword (a sign of war and death).

He goes on to say that ancient civilizations could confirm that comets were ill omens by 'observ[ing] whether disaster did, in fact, follow the appearance of a comet – and it always did'. Lest this sound overly dramatic, Asimov also points out that

> disasters invariably come even when visible comets are *not* present in the sky. The sad fact is that disasters come *every* year and comets have nothing to do with it.[2]

The ancient Greeks viewed comets as being entirely bad, and this view persisted well into the medieval era. To the Greeks, their effects were maximized at about the time of their closest approach to the Sun, although this was more of a general guideline than a hard-and-fast rule. The important thing is that the Greeks felt comets were placed in the sky by the gods and that they were almost invariably portents of doom or ill fortune. Having said that, the early astrologers felt that comets often presaged the birth of a person destined to become important – although, again, this was not necessarily evident until much later, when that person had already risen to prominence.

Modern thinking is somewhat different. No longer do astrologers feel that comets invariably correlate with bad tidings; their new views are somewhat more nuanced. In his online article, Flanery notes that there are different families of comets: those that spend most of their lives in the Oort Cloud, those that spend most of their time in the Kuiper Belt and all of the others. The comets that originate in the Oort Cloud, for example, are thought to be primarily influenced by the stars, and, when they fall into the inner Solar System, bring with them whatever messages the stars have for us on Earth. Kuiper Belt comets, on the other hand, are influenced by the outer planets (Jupiter, Saturn, Uranus and Neptune), and the rest of the comets are influenced by whatever celestial bodies they are closest to during the majority of their orbits. Whether these 'messages' are good or bad, though, depends on a host of other factors, just as with any other astrological portents.

The pre-Christian era

In his book *Fall of a Thousand Suns* (2014), Kevin Curran makes a compelling case that a massive comet struck the Earth about 13,000 years ago, and he speculates that the impact made such an impression on the people alive at the time that it found its way into the religions of the day as well as into the modern era.[3] Curran goes on to discuss the role of comets in ancient religions, including ancient Egypt and India.

First, consider the big picture. As noted earlier, ancient humans studied the sky just as we do today, although for different reasons. Chief among those reasons was to try to glean some sort of understanding as to what they might learn from the stars and planets. If the celestial bodies were put there by the gods – if some of them were, indeed, the gods and goddesses – then any new object that appeared must carry with it some sort of information that could be used to help better understand these same gods and goddesses. Alternatively, the appearance of a comet might be the result of some event in the heavenly realm, perhaps indicating something about the happenings of the gods. Regardless, the appearance of a comet demanded some sort of explanation.

Egyptian mythology, for example, included a story about Apep, a snake-shaped god of chaos, and the underworld. As reported by Curran, sometime around 1100 BCE, Egyptian texts mention the appearance of Apep in the sky; Apep was then chopped into pieces by Seth, the god of storms and defender of the Sun god, Ra. Commenting that the long, bright tail of the sun-grazing comets can stretch across the sky in a narrow band, Curran speculates that this tale could refer to a sun-grazing comet that broke apart when passing near the Sun – something that has been noted on more than one occasion. This could be what the Egyptians saw, but, lacking any context for a natural explanation, they created an explanation that fitted into their existing mythology. Other cultures – Chinese, Native American, Indian and more – came up with similar stories of celestial snakes that were divided into pieces by one means or another. One plausible explanation for these parallel stories is that each of these cultures saw the same thing, explaining it in terms that made sense to them, but constrained by their own views of how the world and the heavens worked. So the Native Americans saw a snake in battle with a bird; Hindus saw the naga (a mythical snake-like creature) and so forth. The point is that the mythology of multiple cultures includes stories and scenes that can plausibly be interpreted as these cultures' attempts to explain a spectacular comet in terms that fitted into their view of the world and the cosmos. Curran makes a compelling case that cometary impacts, near misses and spectacular comet appearances in the sky could have had a profound impact on the religions of antiquity.

Any comet visible to the naked eye will be a surprise to those who are familiar with the skies, because they appear unexpectedly, they follow unusual paths and they change their appearance with time. Comets that grow impressive tails are even more surprising. Such a comet might appear only once in a person's lifetime and can easily be interpreted as having been put in the sky by the gods for some purpose that only they know. Now picture the most impressive comets, the sun-grazers. With a dazzling tail stretching halfway across the sky, these comets are

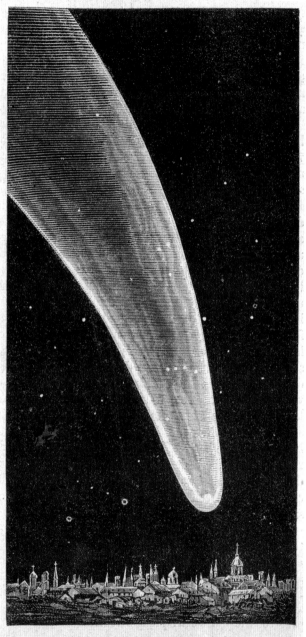

La Comète de 1811.

Engraving from *Le Magasin pittoresque* of the comet of 1811, one of the brightest comets in history. It is easy to see how such a dramatic comet could be mistaken for a god in earlier years.

A painting of Krishna on the serpent Naga, from a Bhagavata Purana manuscript, *c.* 1640.

so impressive that one can easily understand how, especially to a pre-scientific and deeply superstitious populace, such an apparition might be thought to be more than a message from the gods, but a god itself. And it is also easy to understand how such a dramatic vision could be memorialized into religious symbolism – how it could be seen, for example, as a heavenly

95

Etching of a dramatic 1833 meteor shower captured by artist Adolf Vollmy. It is easy to see how such sights could have religious repercussions.

serpent-god stretching across the sky. And when the comet goes behind the Sun, becoming temporarily invisible, only to re-emerge a few days later – sometimes with an additional tail – one can also understand how this could be interpreted as a snake-god losing in battle to, say, a bird-god.

Myths that might be linked to comets abound – far too many to go into detail about all of them here. Let it suffice to say that once you start looking for tales of celestial serpents (or serpent gods), cosmic battles involving serpents and birds, and the visitation of wrack and ruin upon the world from above – any of which Curran suggests might stem from comet sightings or impacts – you see them everywhere. Instances appear in Native American mythology, Norse mythology, ancient Greece, Egypt and more; there are even biblical passages that could easily have been comet-inspired. (As one example, Curran suggests that the story of Satan being cast out of heaven could actually refer to a spectacular comet sighting.)

Humanity's place in the universe

According to the Bible, 'He has fixed the earth firm, immovable' (1 Chronicles 16:30). This is only one of a number of Bible verses that refer to the immobility of the Earth, verses that appear in both the Old and the New Testaments. In this cosmology, the Earth is fixed in the cosmos and circled by the Sun, Moon, planets and stars. This world view became almost as fixed as the positions of the stars, and not without reason – to those whose only tools for observation are their eyes, it seems obvious that this is the case; one need only look around and to the skies to see the truth of that statement. Only when we start looking upwards with better instruments can we see things that make us start to question – perhaps the most famous instance was Galileo's observation that Jupiter had its own satellites, that objects could circle bodies other than Earth. And Galileo's discovery ignited a storm of controversy that continues to this day.

Galileo, of course, did not come up with the idea that the Sun was the centre of the Solar System – that was the work of Nicolaus Copernicus – but he helped to popularize the idea with his book *Dialogo sopra i due massimi sistemi del mondo* (Dialogue Concerning the Two Chief World Systems), and it was this book that got him in trouble with the Church. A lesser-known work of Galileo's, *Il saggiatore* (The Assayer), was ostensibly about comets – specifically, it was Galileo's refutation of an essay on comets published by Orazio Grassi, a Jesuit mathematician. But the book actually dealt as much with the nature of science as it did with comets. (Interestingly, in this case, Galileo was

Justus Sustermans,
Galileo Galilei, 1636,
oil on canvas.

Hieronymus Bosch,
*The Fall of the Rebel
Angels*, from his
triptych, *c.* 1500,
depicting a war in
heaven described
in the Book of
Revelation.

wrong – Grassi ad shown mathematically that comets must lie
beyond the Moon, while Galileo felt that they were an optical
illusion.) Several decades later came the work of Kepler, Halley
and Newton; within a century of Galileo's death, our perception
of the universe changed completely. No longer was the reigning
paradigm that of a universes watched over by an attentive God,
but, rather, it became one of a clockwork universe that was set
in motion by God and that operated according to the fixed
laws of physics. It was the study of comets that helped to
cement Newton's laws of motion and of gravity as the accepted
explanation of how objects in the heavens move.

A bevy of distant galaxies
photographed by the
Hubble Space Telescope.
This photograph contains
images of about 10,000
galaxies up to thirteen
billion light years away
– the motions of the
stars within each galaxy
and the motions of the
galaxies themselves are
governed by the same
laws of physics that
govern the orbits of
comets in our Solar
System.

By helping to demonstrate the efficacy of science in describing the universe, the study of comets helped to force a confrontation between science and religion as two alternate methods for describing the workings of the universe. Three centuries later, although the scientific viewpoint in this particular case (that is, celestial mechanics) has largely prevailed, the underlying controversy continues to wax and wane. Here is why this is significant.

Consider – in a universe that is personally kept in motion by God, anything can happen. If it pleases God suddenly to make the Sun stand still (as the Bible describes in the Book of Joshua), then the Sun will stand still. Events that happen at the whim of God need no further explanation and, by their nature, cannot be predicted. God runs the universe; it is as simple as that. But in a universe that runs according to physical laws, in theory, *everything* can be predicted if only we can learn enough about the universe. Not only that, but if these laws of physics are truly universal, then we, here on Earth, can understand processes taking place anywhere in the universe. In a sense, according to historian of science John Headley Brooke, the development of science that was spearheaded by the work of Kepler, Halley and Newton restored humanity's dominion over nature that had been lost when Man was evicted from Eden.[4]

As important as it was to gain this understanding of the universe and its laws, one can argue that the impact on how humanity saw itself was even more important. Realizing that the universe could be understood was one thing, but realizing that it was people who unravelled these laws was something else entirely. Understanding the universe was suddenly within humanity's grasp. It might take decades (or centuries) to learn the laws of the universe, but people came to believe that it was only a matter of time, a matter of *when* rather than *if*.

This realization has changed humanity's relationship with the universe and, to some extent, with religion. In the pre-Newtonian days, humanity was secure in its position at the centre of the universe, favoured by a God that watched over it and ordered both the heavens and the Earth for humanity's benefit. But ever since that time, humanity has receded increasingly further from the centre of things; astronomically, we have come to realize that we are not even at the centre of the Solar System, let alone the universe. At the same time that our grasp of science and our understanding of the universe exalts our intellect, our increasing distance from the centre of things would seem to diminish our importance *to* the universe. This is part of the religious legacy of the scientific revolution that received so much impetus from the study of comets in the seventeenth and eighteenth centuries.

There is one final point that should be made: believing that an infallible and omnipotent God created and runs the universe is fairly easy to do, and ascribing everything in the universe to the will and the whims of this God is also fairly easy. But what does it say of our confidence in the scientific method – and what does it say about the track record of scientific inquiry – that we are willing to forgo presumably inerrant biblical explanations in favour of the conclusions of entirely fallible people? The history of science is rife with human mistakes, and yet we have come to believe in our ability to understand the universe through the application of scientific principles, primarily because the scientific process, flawed though it sometimes might be, tends to be self-correcting and eventually to arrive at an explanation that will stand the test of time.

Cults

One of the greatest comets of the latter years of the twentieth century was discovered in 1995: Comet Hale–Bopp. As it drew closer to the Sun, astronomers realized it promised to dazzle; this promise was fulfilled in the months to come. More than a year after its discovery, in November 1996, Chuck Shramek, an amateur astronomer in Texas, took a photo of Hale–Bopp; when he looked at the image later that day, he noticed a fuzzy object – somewhat elongated – close to the comet. Intrigued, Shramek compared the object to his star charts, and, when he failed to find a star in the location of the fuzzy object, he contacted a late-night radio talk show host to report what he had found.

What Shramek did not know was that his computer software actually did show a star in the location of his 'discovery'; his computer settings simply prevented it from being shown on screen. So when he went on the air to report his findings, what he said was that he had seen a 'Saturn-like' object (abbreviated SLO) that appeared to be accompanying the comet; this report provoked a number of responses.

The radio programme in question was a late-night show listened to by an eclectic variety of people, especially those who tended to believe in conspiracies, the supernatural and the paranormal. Shramek's report sparked a tremendous amount of interest and follow-up by listeners; it was not long before a self-proclaimed remote viewer announced that she had seen a spaceship accompanying the comet. Another listener, one who claimed to have been contacted by aliens, said that the object was, in reality, a giant planet that would cause cataclysmic changes on Earth when it passed close by. The fact that the fuzzy object seen by Shramek was ultimately shown to be nothing was beside the point; the object had taken on a life of its own by then. These reports played into the beliefs of a cult that had sprung up in San Diego, California, in the 1970s: the Heaven's Gate cult.

One of Heaven's Gate's premises was that humans are able to reach a higher plane of existence; their website, which was still operational as of early 2015, listed a number of biblical verses which they felt supported their beliefs. According to the

The Heaven's Gate logo, from the Heaven's Gate website.

Heaven's Gate beliefs, the only way to reach this higher level of existence is to give up everything on Earth, including their bodies. In their words,

> During a brief window of time, some may wish to follow us. If they do, it will not be easy. The requirement is to not only believe who the Representatives are, but, to do as they and we did. You must leave everything of your humanness behind. This includes the ultimate sacrifice and demonstration of faith – that is, the shedding of your human body . . . In so doing, you will engage a communication of sorts, alerting a spacecraft to your location where you will be picked up after shedding your vehicle, and taken to another world – by members of the Kingdom of Heaven.

A photograph of Comet Hale–Bopp and what was interpreted as a 'companion' object that was, in actuality, a background star. The origin of this image is unknown.

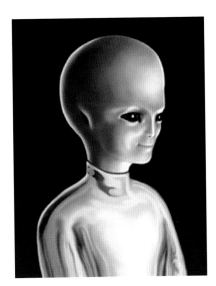

How the Heaven's Gate cult believed a 'Member of the Kingdom of Heaven' might look like.

Reading the Heaven's Gate website and their materials, there is no mention at all of comets, and comets seemed to be no part of their belief system – at least, not until Hale–Bopp appeared in the skies, not until Shramek thought he had found an unaccountable body following the comet during its approach to the Sun and not until the self-proclaimed remote viewer claimed to have seen that the strange body was actually a spaceship. This confluence of events, precipitated by the appearance of a great comet, seems to be what convinced cult members that it was time to move on to the next level in their evolution – by committing mass suicide to free their souls of their bodies. In this, the comet served primarily as a conduit, a messenger, from superior beings to humanity.

What is interesting is that for thousands of years, across an incredible variety of religions and belief systems, and in multiple cultures across the world, comets have come to represent fundamentally the same thing: a messenger (or a message) from the heavens to humanity. They have a host of unusual and dramatic characteristics, and these make it clear to all that they are something extraordinary; it is only natural to wonder why they might have appeared in the heavens, what they are meant to convey to us, whether the news they carry is good or bad, why god sent us this message and for what reason.

5 Comets in Literature and Popular Culture

Comets have inspired not only artists and religious leaders through the ages, but writers, as well as the general public. This inspiration is not limited to the past; comets continue to make an impression on us today, including making frequent appearances in contemporary popular culture.

Although largely confined to the science fiction of recent years, comets were mentioned by Shakespeare in *Julius Caesar* – 'When beggars die there are no comets seen; the heavens themselves blaze forth the death of princes' (Act II, scene 2) – and Seneca wrote of the great comet (mentioned in the previous chapter) that accompanied Caesar's death. Prior to the twentieth century, comets appeared rather sparingly in literature, and tended to be used symbolically. In the last century, comets have appeared more literally as science fiction authors have used them for purposes ranging from brief plot devices to the basis for entire books, including Arthur C. Clarke's *2061: Odyssey Three* (1987).

Comets as signs, portents, omens and so forth

Because of comets' symbolic weight, simply mentioning a comet became a convenient shorthand – a cliché in some sense – for impending disaster and doom, hence the consistent theme in virtually anything comet-related of comets as omens or portents of bad things to come, or as messengers from the gods. We see this theme in the works of Shakespeare, and we see it in Seneca's words proclaiming that the comet in the sky after Caesar's death marked Caesar's apotheosis. In the ages that were ruled by superstition, mysticism and religious belief – in ages in which astrology and alchemy were considered sciences and not belief systems – comets were, of course, viewed through that lens, and the literature of the day reflected these beliefs. The earliest literary mentions of comets are almost entirely supernatural or mystical in nature.

In the Classical era, there were more mentions of comets than Seneca's account of Caesar's comet; one that is mentioned in Sara Schechner's scholarly paper 'Astronomical Imagery in a Passage of Homer' refers to the *Iliad* describing Achilles' preparing to meet the Trojans in battle:

> Next, his high Head the Helmet grac'd; behind
> The sweepy Crest hung floating in the Wind:
> Like the red Star, that from his flaming Hair
> Shakes down Diseases, Pestilence and War;
> So stream'd the golden Honours from his Head,
> Trembled the sparkling Plumes, and the loose Glories shed.[1]

In this verse, the cometary allusion would be the 'sweepy Crest' on Achilles' helmet, likened to 'the red Star, that from his flaming Hair' (comets were sometimes called 'hairy stars' in Classical literature). But Schechner feels that this imagery was actually an insertion by Alexander Pope (1688–1744) in his translation of the *Iliad*; she notes instead that this particular imagery probably originated in a passage from Book Two of *Paradise Lost*:

> Incenst with indignation, Satan stood
> Unterrifi'd and like a Comet burn'd,
> That fires the length of Ophucus huge
> In th' Artick Sky, and from his horrid hair
> Shakes Pestilence and Warr.

Having said that, Schechner acknowledges the presence of a number of other celestial images in the *Iliad*, even if not those

of comets. She also comments on the presence of comet-related imagery by Virgil (70–19 BCE) in the *Aeneid*. Virgil likened the battle-ready Aeneas to both a comet and the Dog Star Sirius:

Achilles dragging Hector's body to the Gates of Troy. Note the flowing plume on Achilles' and Hector's helmets. Franz Matsch, *The Triumph of Achilles*, 1892.

> On the hero's head blazes the helmet-peak, flame streams
> from the crest aloft, and the shield's golden boss spouts
> floods of fire – even as when in the clear night comets
> glow blood-red in baneful wise; or even as fiery Sirius,
> that bearer of drought and pestilence to feeble mortals,
> rises and saddens the sky with baleful light.

Her interpretation of this passage is that Aeneas is 'as fearful as a blazing star and foreboding as the Dog Star ... ready to defeat the ill-fated Turnus and Rutulians in battle.'[2]

In another work, Schechner moves forward in time to the Renaissance with another example of the use of comets in literature, albeit from an era later than Classical times; she notes that the Renaissance poet Torquato Tasso (1544–1595) includes a passage in his *La Gerusalemme liberata* (Jerusalem Delivered) of 1575 (officially published in 1581) that reads

> On his right side he hangs his well-known sword
> Whose seasoned temper flashes recent flames.
> Just as a comet in the burnished air
> Is wont to burn with bloody, horrid locks,

And, wrecking realms, still new disasters bring –
An omen of ill-luck to crimson kings.[3]

Returning to the topic of comets as bad omens, a book written by Thomas Hartman in 1606, *Kometspiegel* (Comet Mirror), lists all of the ills and evils that accompany a comet's appearance:

All comets indeed give evidence
Of a lot of bad luck, affliction, peril and danger
And a comet never appears
To be cares without evil meaning
In general eight kinds of affliction occur
When a comet burns in the air:
Much fever, illness, pestilence and death,
Difficult times, shortages and famine,
Great heat, droughts and infertility,
War, rapine, fires, murder, riots, envy, hatred and strife,
Frost, cold, storms, weather and lack of water,

Sketch to illustrate Torquato Tasso's *Gerusalemme liberata*, showing the comet-like sword, 1590s.

Great increase in people going into decline and death,
Conflagrations and earthquakes in many places,
Great changes in government.

But for us to do penance from the heart, God afflicts us with
disasters and pains.

As Europe reached the point of the Enlightenment, Western
civilization grew less superstitious, and this was reflected in the
literature of the day. In fact, the title of a book published in 1682
by Pierre Bayle is revealing of this tendency: *Letter to M.L.A.D.C.,*
Doctor of the Sorbonne, In Which It Is Proved by Several Reasons
Drawn From Philosophy and Theology That Comets Are Not the

Title page and
frontispiece from
the 1742 edition of
La Lettre sur la comète
de Maupertuis.

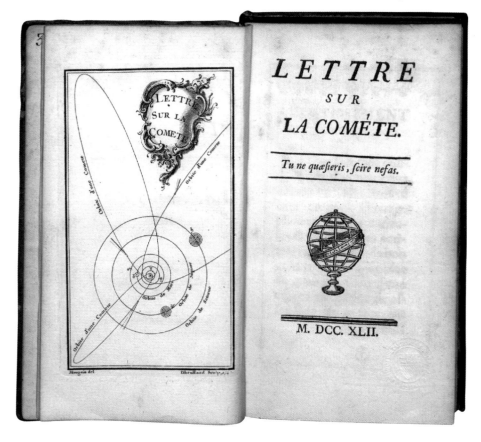

Disasters in Connection with Two Comets Sighted in 1456, from an illuminated manuscript, the Luzerner Schilling, of 1513, showing some of the ills thought to accompany the appearance of comets in the skies.

Presage of Any Misfortune: With Several Moral and Political Reflections and Several Historical Observations and the Refutation of Certain Popular Events. This title tells us that, at this point in time, comets were beginning to lose their ability to induce fear, being seen instead as objects of the natural world. Somewhat later, we find the French author Pierre Louis Moreau de Maupertuis voicing a similar view of comets in his 1742 book *Lettre sur la comète*, in which he noted that comets no longer caused terror when they were seen, that they had 'fallen in such

discredit that they are not thought to be able to cause anything but colds'. But de Maupertuis was a man of the Enlightenment – it is likely that his writing reflected that age's increasing reliance on logic, science and reason as opposed to the superstition and religion that had ruled previous eras.

Bringing this up to the present, we still see the use of comets as bad omens in today's literature, especially in fantasy novels. Perhaps the best-known example of this is in the recent *A Song*

Jupiter, following the impact of a number of fragments of Comet Shoemaker–Levy 9. Impact sites are marked by dark blotches, many of which are larger than the Earth.

of Ice and Fire novels (adapted into the *Game of Thrones* television series), in which a blood-red comet appears in the sky at a critical point, interpreted by many of the characters according to their own background, belief system or needs.

Comets as destructive – and positive – forces

Comets need not be cast earthwards by malevolent deities to be forces of destruction; even in an age devoid of superstition (if such an age ever exists), we still have to acknowledge their capacity for causing harm. For our purposes here, let it suffice to say that we have ample evidence that a miles-wide chunk of ice slamming into the Earth at a speed of thousands of kilometres per hour is going to cause a huge amount of damage. Anyone who saw photos of Comet Shoemaker–Levy 9 striking Jupiter can easily visualize the death and destruction that could ensue on Earth. It should be no surprise that authors, science fiction and otherwise, would have figured this out as well.

Similarly, comets can bring benefits with them without having been sent our way by benevolent beings, though there are fewer examples of this in literature. But this is hardly surprising since most people find it far easier to view their destructive potential than the possibility of good. Nevertheless, comets do show a positive face from time to time, and this warrants some discussion as well.

First, it can be difficult – especially in earlier literature – to differentiate between comets acting as natural phenomena and those acting as supernatural agents. In fact, not until the latter part of the eighteenth century can we really say that the idea of comets as a source of purely natural devastation came up. This was in a scientific paper, *Réflexions sur les comètes qui peuvent approcher de la terre* (Reflections on the Comets that Could Approach the Earth), by the French astronomer Joseph Jérôme Lefrançois de Lalande in a presentation to the French Academy of Sciences. Lalande's presentation was a piece of scientific research; among the reactions it sparked was Voltaire's *Lettre sur la prétendue comète* (Letter about the Alleged Comet) of 1773, in which the French writer mocked the idea that a comet would strike the Earth on a particular day, causing terrible harm. Although in his arguments Voltaire proved himself a better satirist than scientist, he was correct in his assertion that a predicted cometary impact would not cause the foreseen destruction, presaging similar arguments we still see today.

Pen and ink illustration of *The Conversation of Eiros and Charmion*, by Henry Clarke (1839).

Slightly more than a half-century later, Edgar Allan Poe (1809–1849) wrote what could be called one of the first end-of-the-world science fiction stories, *The Conversation of Eiros and Charmion*, in which two departed souls (the eponymous Eiros and Charmion) discuss the destruction wrought on Earth by a comet. While acknowledging that comets are not very substantial ('We had long regarded the wanderers as vapory creations of inconceivable tenuity, and as altogether incapable of doing injury to our substantial globe, even in the event of contact'), they saw the Earth fall prey to chemistry when a comet somehow stripped the nitrogen from the Earth's atmosphere, leaving it composed of almost entirely oxygen:

That tenuity in the comet which had previously inspired us with hope, was now the source of the bitterness of despair. In its impalpable gaseous character we clearly perceived the consummation of Fate. Meantime a day again passed – bearing away with it the last shadow of Hope. We gasped in the rapid modification of the air. The red blood bounded tumultuously through its strict channels. A furious delirium possessed all men; and, with arms rigidly outstretched towards the threatening heavens, they trembled and shrieked aloud. But the nucleus of the destroyer was now upon us; – even here in Aidenn, I shudder while I speak. Let me be brief – brief as the ruin that overwhelmed. For a moment there was a wild lurid light alone, visiting and penetrating all things. Then – let us bow down, Charmion, before the excessive majesty of the great God! – then, there came a shouting and pervading sound, as if from the mouth itself of HIM; while the whole incumbent mass of ether in which we existed, burst at once into a species of intense flame, for whose

"THE BEST END-OF-THE-WORLD STORY SINCE
ON THE BEACH—SUPERB DETAIL, SHUDDERINGLY
BELIEVABLE."—FRANK HERBERT,
AUTHOR OF DUNE AND CHILDREN OF DUNE

LUCIFER'S
HAMMER

LARRY NIVEN AND JERRY POURNELLE

2-3599-8•$2.50

FAWCETT
CREST

Cover of the comet
disaster novel *Lucifer's
Hammer* (1977).

surpassing brilliancy and all-fervid heat even the angels in
the high Heaven of pure knowledge have no name. Thus
ended all.

Several decades later, the French astronomer Camille
Flammarion (1842–1925) wrote a novel, *La Fin du monde* (The
End of the World), which began with a comet striking the Earth,
with disastrous results. Flammarion writes about the response of
the public and the scientists to news of the upcoming inevitable cometary impact; astronomers wrote somewhat dispas-
sionately about the science behind the upcoming extinction of
humanity, while the news largely passed by the public or was
received with disbelief. As Flammarion wrote,

> Neither the first announcement of the press, that a comet
> was approaching with a high velocity and would collide with
> the Earth at a date already determined; nor the second, that
> the wandering star might bring about a general catastrophe
> by rendering the atmosphere irrespirable, had produced the
> slightest impression; this two-fold prophecy, if noticed at all
> by the heedless reader, had been received with profound
> incredulity . . .
>   That a collision with the earth would occur was certain
> . . . That was a fact which mathematics had rendered certain.
> The absorbing question now was the chemical constitution
> of the comet. If the Earth, in its passage through it, was to
> lose the oxygen of its atmosphere, death by asphyxia was
> inevitable; if, on the other hand, the nitrogen was to com-
> bine with the cometary gases, death was still certain; but
> death preceded by an ungovernable exhilaration, a sort of
> universal intoxication . . . .[4]

In Flammarion's book, the comet brings about the end of
humanity. A half-century later, the Finnish author Tove Jansson
(1914–2001) developed an elaborate world of Moomins (somewhat bohemian troll-like creatures) that featured in a number of
her books and comic strips. In the book *Comet in Moominland*

of 1946, Jansson writes about a comet that is predicted to strike the planet; some of the Moomins experience drought, streams and seas drying up, and are faced with the heat from the comet on their way to seek shelter. They emerge from the cave in which they have been hiding to find that the comet missed them after all. In this case, Jansson might have been using the comet as a metaphor for nuclear weapons, which had recently been developed, deployed and become public knowledge.

The 'comet as destroyer' genre reached its pinnacle with the book *Lucifer's Hammer* of 1977 by the science fiction greats Larry Niven and Jerry Pournelle. One of the classics of science fiction, *Lucifer's Hammer* is possibly the definitive comet impact novel, tracing events from the discovery of a giant comet through to its impact, the fall of much of society and the attempts of survivors to reconstitute a degree of civilization.

Interestingly, over more than a century, the 'comet as destroyer' theme has focused not so much on the science as on the societal impact (if you'll forgive the pun) of a comet-caused disaster. Larry Niven is renowned for his mastery of 'hard' science fiction, a subgenre which focuses on the realistic and technical aspects of science and technology, and Flammarion was, in fact, a scientist. Yet even they spent the majority of their books discussing how human society was affected by the comet's impact. In these fictional stories, the comet was really not much more than the vehicle used by these authors to explore issues surrounding so huge a stress to society. In fact, in their writing, the science – however fascinating – is not nearly as interesting as how the authors speculate people and institutions will respond to such a stress, a theme that we see in the popular zombie apocalypse television series *The Walking Dead*, in which the writers use a scientific problem to explore how society would respond to a near-apocalyptic disaster. Is it better, for example, to be led by a strong, self-appointed leader or to vote for a leader? Is it better to follow as closely as possible the same ethical code we have today, or do people do better by reverting to an every-man-for-himself ethos? Fiction like this even touches upon something as simple as whether or not a technological civilization can survive the loss of so many people and so much infrastructure. These questions were interesting in and of themselves a century ago, before we had any reason to believe that such impacts could happen. Today, humanity has developed the ability to wreak havoc with our own weapons, not to mention our understanding of what a giant impact can do to our planet; this gives today's end-of-the-world speculations a relevance and a place in the public consciousness that was lacking in the years before the first nuclear explosion.

Having said all this, some comet stories have happy endings. In Kim Stanley Robinson's *Mars* series, comets are diverted into the planet to bring water for terraforming. A race of humans called Conjoiners in Alastair Reynolds' *Revelation Space* series use comets as homes, burrowing into the ice in the Oort Cloud to make habitats. And Dan Simmons's *Hyperion* series also features a branch of humanity that uses comets; in this case, the branch lives in the deep space between the stars and the inhabitants are called Ousters – they use comets to, in effect, water their habitats. In these books, the comets are used as a natural resource and nothing more, bringing water (and sometimes more complex chemicals) to barren worlds. Only H. G. Wells (1866–1946) seems to have portrayed comets as beneficial in ways other than as a resource: in *In the Days of the Comet* of 1906, Wells postulates a comet that enters the Earth's atmosphere and fills it with a green gas; the gas puts everyone to sleep and changes them, making them more rational and altering the way they view themselves, others and the world.[5] In this case, humanity is the beneficiary of what the comet brings to the Earth.

So there are a number of stories that show comets in a positive light, but not many – when it comes to literature, it turns out that comets are more likely to be seen as being forces of destruction.

Everything we have talked about thus far is about comets, but more about what comets portray and what they can do than about the comets themselves. But after hearing about what comets can do (good and bad) or what they might portend, it is

JULES VERNE
HECTOR SERVADAC
DESSINS
DE
P. PHILIPPOTEAUX
GRAVÉS PAR C. LAPLANTE

Paul Dominique
Philippoteaux's
illustration from the
1877 edition of Jules
Verne's book *Hector
Servadac's Voyages and
Adventures Across the
Solar System.*

only natural to wonder if anything is ever written about the comets themselves – not as agents of the gods or as forces of nature, but as objects in and of themselves. And it turns out, there is. In particular, there are a number of books and stories that are all about the comet; unsurprisingly, these tend to be hard science fiction stories.

Two early stories both use comets as vehicles. Jules Verne's (1828–1905) *Hector Servadac, voyages et aventures à travers le monde solaire* (Hector Servadac's Voyages and Adventures Across the Solar System, 1877) tells of the comet Gallia, which strikes the Earth a glancing blow and scrapes off a chunk of our planet, including 36 assorted Europeans and the occasional animal. The comet carries the group around the Solar System while they try to figure out what has happened and explore their new world – they are finally able to transfer back to the Earth when the comet returns after two years. Another cometary travel story was written only a few decades later by the American author Mark Twain (1835–1910) in his story *Captain Stormfield's Visit to Heaven,* of 1907. In this story, told with typical Twain humour, the title character hitches a ride on a comet to reach heaven after his death. Part of the story (the lesser part, although highly entertaining) describes the route of the comet and likens it to a nineteenth-century sailing ship, this one making deliveries throughout the cosmos, including to heaven and hell. The opening lines nicely sum it up:

> Well, when I had been dead about thirty years I begun to get a little anxious. Mind you, had been whizzing through space all that time, like a comet. Like a comet! . . . But after I got outside of our astronomical system, I used to flush a comet occasionally that was something like. We haven't got any such comets – ours don't begin. One night I was swinging along at a good round gait, everything taut and trim, and the wind in my favor – I judged I was going about a million miles a minute – it might have been more, it couldn't have been less – when I flushed a most uncommonly big one . . . The comet was burning blue in the distance, like a sickly

torch, when I first sighted him, but he begun to grow bigger and bigger as I crept up on him. I slipped up on him so fast that when I had gone about 150,000,000 miles I was close enough to be swallowed up in the phosphorescent glory of his wake, and I couldn't see anything for the glare.

Another version of the comet-as-vehicle line is *Heart of the Comet*, by science fiction authors David Brin and Gregory Benford. Here, astronauts are dispatched to Halley's Comet on one of its trips through the inner Solar System; the goal of the mission is to learn enough to divert the comet into a closer orbit so that it can be harvested for its resources. What the crew finds is that the comet is teeming with life; when their spaceship is disabled, they get to experience all that life up close and personal. The book revolves around the crew's struggle to survive until the comet makes its next trip back to the inner Solar System; along the way, they have to fight off invasive organisms, disease, starvation and more. They also have to decide whether they are going to merge with some of the organisms on the comet or to stay genetically intact. In fact, the crew ends up breaking into two factions, those who do merge and those who refuse to do so, and these factions end up embattled through parts of the book. *Heart of the Comet*, released to coincide with Halley's visitation in 1986, was the first mainstream science fiction novel actually to be set primarily on

Cover of Mark Twain's *Captain Stormfield's Visit to Heaven* (1909).

Frontispiece of *Captain Stormfield's Visit to Heaven*, showing Captain Stormfield riding the comet.

Logo of the Utica Comets, a hockey team based in the small town of Utica, New York.

(and inside) a comet. And although much of the story revolves around the science and the crew, the book also touches on topics such as genetic engineering, immortality and even what it means to be human.

Although there have been a number of comet-based science fiction stories, one in particular bears mention to close this section: the novel *Pushing Ice* of 2005 by Alastair Reynolds. This story is set on a spaceship, *Rockhopper*, designed to divert comets from the outer Solar System and send them inwards to be captured and used for their resources. But when Saturn's moon Janus suddenly starts moving out of the Solar System, *Rockhopper* is dispatched to find out why. What ensues is a huge and complex story that takes readers across the universe and through immense amounts of time. And although comets play a very small role beyond the first section of the book, they provide a plausible reason – one that is under serious consideration today – for a ship to be in position to rendezvous with Janus, not to mention that Reynolds does a laudable job of getting the science right and was complimented on this by the Astronomical Society of the Pacific (a respected organization of professional astronomers and educators).

What all of these stories have in common is that they are set – in part or in whole – on comets, and the comets play a significant role in the story. Whether humans are hitching a ride, investigating the comet or trying to harvest it, these stories use science as more than just a vehicle to introduce other topics. In most, the comet is integral to the story itself, and in the hard science fiction stories, the authors make a serious effort to avoid scientific mistakes. Even if the comet is the means to an end (that is, comet mining as a reason for spaceships to be in the vicinity of Saturn), the science can be an important part of the story, and the writers treat the comets as objects worth more than a passing mention or description.

While comets have been mentioned in literature for millennia, they have been in popular culture for only a century or so. Examples abound of comets in our popular culture: Captain Kirk hid his starship behind a comet in the original *Star Trek* series; a comet struck the Earth in the movie *Deep Impact* of 1998. Comets even appeared in an episode of the popular television shows *Friends* and *The Simpsons*. There are also sports teams, consumer products and more that are named for comets. Many scientists continue to appear in popular culture as well, and this can be a point of national pride, although it might be stretching the definition of 'popular culture' somewhat to include the portrayal of Isaac Newton on the £1 note (now out of circulation) in the category. Having said that, money is something that most people see and handle daily.

Popular science: writing, news and documentaries

Non-fiction books and articles are simply too numerous to provide even a partial listing here; there have been hundreds of books and thousands of magazine articles written about comets in the last several decades and likely tens or even hundreds of thousands of newspaper articles. The reason for this, quite simply, is that the public seems to be fascinated with comets, and scientists keep learning more about them. There have been a number of recent missions to study comets, and each mission grows in complexity and drama – as well as yielding new discoveries and a better understanding of comets with each of these missions. Between the public's interest in a beautiful phenomenon and the inherent drama of the subject (a rendezvous with a comet, smashing a probe into a comet, landing on a comet, for example) – along with the continuing scientific interest in continuing to study these objects – there are seemingly endless opportunities to continue writing about the science of comets. Television documentaries and extended television and radio stories about comets are driven largely by the same factors: breaking news stories about new scientific studies, the return (or initial appearance) of comets both major and minor, and new and more daring space missions. Add to that the hundreds of recurring comets and the (most likely) trillions of comets in the Oort Cloud (most of which will be written about when they descend into the inner Solar System), and it seems likely that science writers and scientists alike will continue to write about comets for decades or centuries to come.

Artist's conception of the Deep Space 1 mission's approach to Comet Borrelly.

Detail of a one-pound note featuring a portrait of Isaac Newton and a representation of the orbits his physics helped to describe.

To this, we can add the news coverage of the comets that appear in the skies and of our cometary science missions. People need not be science geeks to take an interest in the subject – people follow the news about comets because comets are beautiful, because they are rare (especially the great comets), because they can pose a threat to Earth and for any number of other reasons.

The broad coverage of comets in the news began in 1973, when the world's attention turned to Comet Kohoutek, whose rather disappointing appearance belied its initial promise to be spectacular. Kohoutek also made an appearance in the *Peanuts* comic strip and it was featured in a number of songs of the 1970s and 1980s, such as Kraftwerk's 'Kometenmelodie' from the

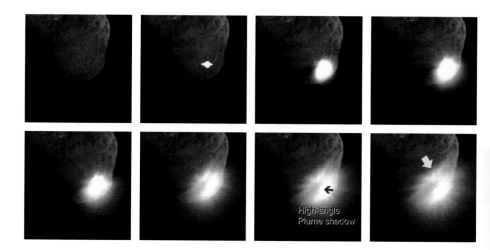

album *Autobahn* (1974) or Journey's single 'Kahoutek' from their debut album of 1975, in addition to playing a role in inspiring the best-selling book *Lucifer's Hammer.*

Somewhat over a decade later, another great comet, Comet Halley, was met by a flotilla of space probes during its visit to the inner Solar System in 1986. Like Kohoutek, Halley received a tremendous amount of public attention, including in books, television shows and coverage in all of the media. And also like Kohoutek, Halley was something of a disappointment when it finally appeared in the sky.

In the following decade, public attention was drawn to yet another comet, Comet Shoemaker–Levy 9, which broke up into 21 pieces and hit Jupiter in 1994, driving home the destructive power of celestial impacts. The impacts became among the most-photographed events in the Solar System, and images made their way into the world's media, including onto the then-new World Wide Web. Just a few years later, yet another comet entered the public consciousness when the spectacular Comet Hale–Bopp came into view.

The next major comets to enter the public awareness were not even visible to the naked eye, but they still captured the world's attention; these were comets visited by spacecraft that did their work under the scrutiny of the global press. The first

Photo sequence showing Comet Tempel 1 during NASA's Deep Impact mission just before, during and after being struck at high velocity by a probe released by the spacecraft. The arrows indicate various shadows cast by the ejecta.

of these was Comet Tempel 1, which was struck with a projectile during the Deep Impact mission of 2005 to study the comet's composition. Just a year later, Comet Wild 2 had some of its tail trapped and returned to Earth in 2006 as part of the NASA Stardust mission; and in 2014 Comet 67P/Churyumov–Gerasimenko hosted the first soft landing of a space probe followed by the beaming back of photos from the comet's surface.

The moment of impact; probe strikes comet at 5.45 am (Universal Time) on 4 July 2005.

Popular fiction and modern fantasy

Comets show up repeatedly as elements or as themes in comic books and graphic novels, in popular novels and so forth. Not only that, but there is a realm of science writing (this book, for example) that bring comets to the general public. Some of the available writing is book-length, more is article-length, but it is all aimed at the interested public. The majority of the writing is science fiction and, in it, comets tend to be used, one might say, literally – as comets. They might be the entire focus of the plot or a major plot point, or they might be mentioned only in passing; the point is that when comets appear in science fiction stories, they are most likely there representing nothing other than themselves – in other words, science fiction is typically interested in the science of comets or in comets as physical objects as opposed to using them symbolically in some way.

By comparison, mainstream fiction and fantasy novels might make use of comets as literal objects, but more likely is that they will be included for their symbolism. One of the more interesting symbolic uses of comets was in George R. R. Martin's *Song of Ice and Fire* series of novels, as mentioned earlier, with the comet meaning something different to every culture that witnesses it; in such a complex story, the comet might well have been imbued with more symbolism than is fair for any single astronomical object. A single comet in this series, for example, was interpreted as predicting the coming of dragons, war, vengeance, a new king or a new age to come. Others saw it as a portent for Daenerys (one of many attempting to claim the Iron throne) and her followers to guide them through the Red Waste desert region of Essos in her quest to return to the continent of Westeros, to rule her ancestral lands; some felt that its appearance was an omen that House Lannister (one of the major forces in this series) will come to power; and still others saw just the opposite – as an omen that the forces of Robb Stark (leading another of the major powers) would defeat its enemy, House Lannister.

These cultures not only had their own interpretations of the comet's meaning, but each had its own name for the comet.

Some of these names were symbolic (The Sword that Slays the Season, The Red Messenger, The Bleeding Star); others reflected the powerful people of the story (Lord Mormont's Torch, King Joffrey's Comet); still others were based on mythologies (Dragon's Tail, The Red Sword). What it boils down to is that the people – the cultures – in this series see in the comet whatever they are predisposed to see, the same thing that has happened throughout human history with virtually every culture on Earth. The comet is a Rorschach test for the inhabitants of the world, or at least for the characters of the story.

Television shows and movies

Comets have appeared across the spectrum of genres in television and the movies. In the various Star Trek franchises, for example, Captain Kirk used a comet to hide his battered *Enterprise* from Romulan enemies, while in the prequel series *Star Trek Enterprise*, a shuttlecraft evades detection by hiding behind a comet that has been diverted to strike Mars as part of a terraforming project. Comets were used as a weapon in a *Doctor Who* episode, when the evil Cybermen planned to steer a comet into Earth to cause widespread destruction, while the villain Darkseid diverted 'Fleischer's Comet' in the cartoon *Superman: The Animated Series* to create havoc on Earth. Other comets that threatened doom to the Earth appeared in the movie *Deep Impact*, in which a comet 7 miles (11 km) in diameter threatens to cause a mass extinction on our planet; in the television movie *A Fire in the Sky*, when a comet strikes the city of Phoenix, Arizona; and in an episode of *The Simpsons*, when Bart discovers a comet that is projected to strike Springfield (in another episode, Homer claims to remember when Halley's Comet hit the Moon).

In a different animated series, *Futurama*, the characters reference a war to reclaim Halley's Comet, and in another episode, the characters mention that Halley's Comet had been mined for all of its water. And while a comet did not actually strike the Earth in the disaster film *The Night of the Comet*, its visit turned people into zombies with the expected bad results. Moving away

from science fiction, the 1987 movie *Roxanne* (based on the legend of Cyrano de Bergerac) features a lovely graduate student, whose thesis project is based on predicting a comet's return, being wooed by the main character. Still another television series, *Millennium,* made occasional mention of a double-tailed comet as a portent of bad things to come – in this case, it turned out to be a deadly plague. Finally, there was an episode of the popular series *Friends*, in which main character Ross Geller takes his friends to the roof to watch the appearance of a comet.

As in so many other areas of our culture (popular and otherwise), comets have come to be used in any number of ways. They can be used as comets – as in many science fiction stories – but they can also be used symbolically, as they are in so many other cultural areas. But television and movies are visual media; they also make use of comets simply because they look beautiful and dramatic. For visual media, comets are perfect.

Halley's Comet only appears about once every human lifetime, as shown in this poignant cartoon by the Malaysian artist Tan Yong Lin.

Popular art and graphics

The visual beauty of comets also shows up in popular art and graphics – not only in paintings, but in comic strips, cartoon shows, anime and manga, graphic novels and so forth. Part of this is because these often have a science fiction theme – either explicitly or implied – and part, as mentioned above, is simply because they are such beautiful objects. In many cases, comets are used as visual clichés; they are included to indicate that a particular cartoon is set in the future or in space. Like ringed planets, comets might have absolutely nothing to do with a story, but showing them makes it clear that the story is set in space or the events are taking place in the future.

Popular art, of course, can also include editorial cartoons. These might poke gentle fun at worries of global calamity, they can be commentaries on the space programme or they can simply provide a humorous view of the latest discoveries from the ground or in space. No matter how they are intended, these cartoons capture society's view (or at least that of the cartoonist) at the moment they are written. They are not meant to reflect deep thought about the future of humanity, our place in the universe, the philosophy of scientific exploration of the universe and so forth so much as comment on the thoughts of the moment – the zeitgeist, if you will.

Popular art also makes extensive use of photographs of comets, often manipulating the photographic images for artistic effect. Astronomers frequently use computers to process digital images to help analyse their photos and to bring out details of scientific importance; artists will process photos as well, either digitally or by adjusting the colours and chemistry when working with film, to bring out artistic details or to help (or force) the viewer to see the comets in – quite literally – a different light. This might involve cropping photos to eliminate distracting details, changing the colours to a more dramatic palette or sometimes simply narrowing the palette of colours to force us to see a simplified image, perhaps one that is focused on the shapes or the relationship of the comet to the background and (ultimately) to the viewer. Even if we know what the reality looks like – even if we can see the unadulterated image of a comet in our mind's eye – we are forced to see the image, the colours and the shapes, put in front of us by the photographer and to acknowledge the photographer's view of, in this case, the comet.

Comets in games

Computer games are widespread, and a large percentage of these are in the genre of fantasy or science fiction, including a number that have comets as part of the plot lines or as dramatic graphics. These include *Jesus: Dreadful Bio-monster* (1987), in which Halley's Comet is found to harbour life during a close approach to Earth, and *Shadow of the Comet* (1993), in which certain game entities can only be called forth during the passing of a particular comet. There is also *Super Mario Galaxy* (2007), in which a comet-like structure named the Comet Observatory is used by a main character to travel through space, and *Final*

Woodcut of an astronomer observing a comet. Illustrated by an artist who goes by the name of Xunantunich.

Fantasy VII, in which a comet can be called forth by a malevolent character to wreak havoc in the galaxy, finally impacting the Sun and causing massive damage. Comets are also used maliciously in *Shadow the Hedgehog* (the Black Comet not only holds most of the game's enemies, but spreads a paralysing gas on the planet), but more equivocally in *Illusion of Gaia*, in which it is explained that comets are used by ancient civilizations to speed humanity's evolution, but were turned into weapons. Finally, there are those games where comets appear as omens: in *Warhammer Fantasy Roleplay*, a two-tailed comet is an omen of great change and has a prominent religious symbolism in the game's 'universe', and in *Myth*, a comet's appearance every millennium signals the transition from a world of light to a world of dark, or vice versa.

Editorial cartoon from the 8 July 1881 issue of the San Francisco-based magazine *The Wasp*: railroad tycoons Leland Stanford and Collis Potter Huntington, while robbing graves, are about to be struck by a comet labelled 'Retribution'.

To all of this we can also add products such as Comet brand cleanser and Comet paper plates. Comets have also made an appearance in popular music as the subject of song (Mary Chapin Carpenter's song 'When Halley Came to Jackson' tells the story of a child born during the comet's appearance in 1910 and who lives to see it return in 1986), as well as in the name of the 1950s musical group Bill Haley and his Comets.

The fact that we have popular television shows in which comets play a role, that there is a kitchen-cleaning product with the word 'comet' in its name, that we have editorial cartoons featuring comets that run in major newspapers and magazines and that even children's games (and those of adults as well) feature comets shows us how deeply these objects have permeated our culture.

6 History's Greatest Comets

There are hundreds of billions, possibly trillions, of comets in the Solar System, and, on average, there are one or two each year that make it close enough to the Sun to be seen with the naked eye. But most of these are unremarkable – they are usually very close to the limit of what we can see with the naked eye, and, if anything, they are likely to look like a small, faint fuzzy spot in the sky. Some comets never get close enough to the Sun to sprout a tail; others, having made too many trips to the inner Solar System, simply do not have enough dust left to form a tail. But even if a comet does grow a tail, it might not be at all remarkable – there has to be the right combination of high levels of dust, a close passage by the Sun and a position in the sky where it can be seen if a comet is going to catch our attention. So while there might be comets showing up every year or so (sometimes more – there were five naked-eye comets in 2004!), most of them come and go without making a splash outside the community of professional and amateur astronomers.

Every now and again, there will be a bright comet with a bright tail – something that everyone can see and that captures the eye. These comets are in the sky every decade or so, and anyone seeing them will almost certainly carry the memory for a lifetime. These are the comets that look like our conception of comets: a bright nucleus surrounded by a dim and fuzzy coma, with a dramatic tail sweeping across the sky. But this chapter is about those comets that are so dramatic and so awe-inspiring that they are remembered for centuries or even millennia; some of them even changed history. The science author D. Justin Schove

An 1857 pamphlet announcing the forthcoming appearance of a comet.

developed a scale by which comets can be rated.[1] The scale ranged from one (minor comet, noted only by experienced skywatchers) through nine (created terror, remembered for generations), depending on their appearance in the sky, as well as their impact on the society of the time.

Comets of the first three levels would be most likely to go unnoticed by anyone except the most dedicated naked-eye sky-watchers – the astronomers of ancient China come to mind.

Today, armed with sensitive cameras, computer software and a detailed knowledge of the sky, such comets are far more likely to be noticed by both amateur and professional astronomers.

The middle part of this scale includes the run-of-the-mill comets – apparitions that might be noticed by an ordinary person but might not warrant more than passing attention. These are the comets that could be mentioned in the newspaper (but back in the science section), and might even garner a brief story on the nightly news. But these, too, will fade from memory with time; they certainly will not make history.

The truly great comets are the ones that demand to be noticed. They blaze in the sky for weeks or months; they might even be visible during the day. In ages past, these comets might serve as portents of war or might be thought to mark the passing of kings; these are the comets that are remembered by history – the comets that might even change history. In this chapter, we are going to be talking about comets at the top end of the scale.

One thing to remember is that through most of history, comets shared the sky only with the stars, the planets and the Moon – there were no streetlights or headlights, no spotlights shining into the night sky, no billboards or advertisements – nothing to distract the eye or to compete with whatever nature put in the heavens. Not only that, but there were fewer distractions inside as well; those who stayed up after dark were not distracted by the television or the radio, and, with most of the population illiterate through most of human history, there was little competition for a person's attention. Barring bad weather, whatever appeared in the sky was bound to be seen by whoever took the time to look, and in the completely black skies, any great comet that did appear was bound to look far more spectacular than it would today.

One final note regarding any list of any astronomical bodies is going to be heavily weighted towards recent years. For faint objects (asteroids or distant supernovae, for example), this is simply because we have more telescopes viewing the heavens with better equipment than ever before – of course we are going to see more! But the truly great comets do not require sophisticated instruments to be seen; if they did, the term 'great' simply would not apply. And it is not likely that we happen to live in a particularly comet-rich time in history. More likely, according to author David Seargent, is that the southern hemisphere is more heavily populated and more advanced today than at any time in the past – there are more people watching the skies, and it is easier for them to communicate what they see to the rest of the world.[2] We can only see part of the sky from the northern hemisphere, while comets can appear in any part of the sky following orbits with any inclination; Seargent points out that a comet might be missed if it reaches its maximum brightness or makes its most dramatic appearance in a part of the world where it is either not seen or where there are no permanent records made of its appearance – historically the southern continents were populated, of course, but in large part by societies that did not keep written records. No matter how great a comet might have been, we cannot know about it if it went unnoticed or unrecorded.

Comet Halley (multiple appearances)

Mark Twain was born when Halley's Comet (technically it should be called Comet Halley, but it is popularly known as Halley's Comet) was in the sky, and it returned at the time of his death. Halley's Comet is woven into the Bayeux Tapestry depicting the Norman invasion of England in 1066, and nearly a millennium later, it became the first comet ever visited by spacecraft. Halley's Comet was the first whose return was scientifically predicted, and astronomers, using their knowledge of its orbit, have linked it to records of comet appearances as far back as 250 BCE (Greek and Chinese astronomers reported a comet that appeared in 467 BCE, which was likely Halley's Comet as well, but the dates are uncertain and the identification is not definitive). Anybody in the world, if asked to name a comet, will likely name this one. Halley's Comet has not only had a significant cultural impact, but has had a profound scientific impact as well – for these reasons (and more), it is very likely the most significant comet in history.

Page from the
Eadwine Psalter
(1120 CE), featuring
a sketch of Halley's
Comet at the bottom.

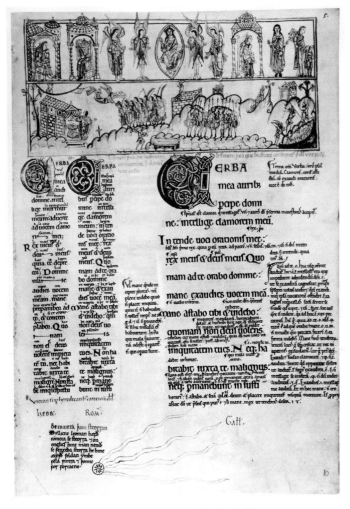

On the scientific side of things, Halley's Comet is not a very large object – although the coma can spread out to 100,000 km or more, the nucleus is more or less a potato-shaped oblong only about 15 × 8 km (9 × 5 miles). When the Giotto space probe closed on the comet in 1986, its images confirmed Whipple's 'dirty snowball' hypothesis of comet composition and structure (more or less). The comet spends most of its time in the outer Solar System, in the realm of the gas giants and beyond. At its greatest distance, Halley's Comet is as much as 35 times as far from the

Sun as is the Earth (as noted earlier, Earth is at a distance of 1 astronomical unit, or AU, from the Sun) – somewhere in the vicinity of Pluto's orbit. At its closest, it ends up between the orbits of Mercury and Venus (about 0.6 AU). Halley's orbit is highly elongated – in astronomical parlance, it has an eccentricity of over 0.95 (the eccentricity of a perfect circle is 0, and an eccentricity of 1 means that an object will never return) and is inclined at 18° to the ecliptic. Unusually, Halley's circles the Sun in the opposite direction from the planets and asteroids – this is called a retrograde orbit – completing a circuit about once a lifetime (about 76 years). Interestingly, its orbital period has changed over the centuries, most likely due to the influence of Jupiter's gravity tugging on the comet during close encounters, along with the action of jets of gas that push it somewhat off course when it approaches the Sun.

Like all comets, Halley's formed from the primordial Solar nebula and is among the most ancient objects in the Solar System. After spending nearly five billion years quietly circling the Sun in the Oort Cloud, about a hundred millennia ago, something jarred it loose and it fell towards the Sun. In the outer Solar System, it was in the deep freeze, circling nearly unchanged for aeons. Once it began passing close to the Sun, it started to change. Astronomers estimate that it has lost over 80 per cent of its original mass during its trips through the inner Solar System, and it will most likely not last more than another few tens of thousands of years. Until then, it will continue lighting the skies every time it drops into our part of the Solar System.

One of the reasons that Halley's Comet so captures our attention is that it passes very close to the Earth's orbit, so close in fact that we are periodically pelted with its debris in the form of meteor showers – specifically, the Orionids that come along every October, when the Earth passes through Halley's orbit. Even more interesting than this, though, was that in the appearance of Halley's Comet in 1910, the Earth passed through the comet's tail. This appearance was heavily analysed by astronomers, and Halley's Comet became the first comet ever to have its gases analysed by spectroscopic analysis, revealing the presence of cyanogen. As mentioned earlier, this caused a degree of panic.[3]

The appearance in 1910 was one of the closest approaches to Earth in centuries, and it was spectacular in the skies. In addition to the panic mentioned in the previous paragraph, it also contributed to the civil unrest that already permeated China; while the comet was not responsible for the unrest, China scholar James Hutson reported,

> A comet is a very unlucky omen; and the appearance of Halley's comet in 1910–11 brought with it a great deal of unrest and fear. The people believe that it indicates calamity such as war, fire, pestilence, and a change of dynasty. In some places on certain days the doors were unopened for half a day, no water was carried and many did not even drink water as it was rumoured that pestilential vapour was being poured down upon the earth from the comet.

Halley's appearance in 1986, by comparison, was a scientific bonanza but a visual dud. Not only was it three times as far from Earth as it was during the appearance in 1910 (making it only about one-tenth as bright), but the widespread light pollution of the late twentieth century meant that the skies in 1986 were far less conducive to good viewing than were the much darker skies of 1910. The comet is due back again in 2061, when it should be one of the brightest objects in the sky.

372 BCE (Aristotle's Comet)

By all accounts, the great comet of 372 BCE was an amazing sight – bright enough to cast a shadow at night with a huge tail stretching nearly halfway across the sky. In later centuries, writers, repeating stories now lost, claimed that the comet broke up into multiple pieces. If so, this would be the earliest documentation of such an event, a phenomenon similar to the fate that befell the recent Comet Shoemaker–Levy 9 before it plunged into Jupiter's atmosphere in 1994.

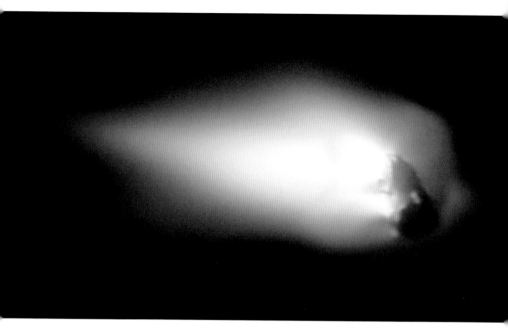

From the descriptions of Aristotle's Comet – a very small nucleus with a huge, bright tail – it might well have been in the class of comets known as Kreuz sun-grazers, comets that barely miss the Sun at their closest approach. These are among the most dramatic of the comets; their close approach to the Sun makes them among the brightest objects in the sky as they whip past it, moving faster than virtually any other body in the Solar System. These objects capture the attention of anyone who can see them – this particular comet reportedly was so bright as to cast shadows.

A view of the nucleus of Halley's Comet as photographed by the ESO's Giotto spacecraft during its 1986 rendezvous.

In his book *The Greatest Comets in History*, author David Seargent notes the comet's possible breakup and suggests that Aristotle's Comet might be the progenitor of some or even all of the sun-grazing comets.[4] Without accurate observations, there is no way to know the orbit of this comet, so there is no way to know whether or not this is the case. We do know that there are a number of these sun-grazing comets that seem to follow similar orbits, and it seems reasonable to assume that they originated from a parent comet that broke up; it is certainly plausible to suppose that many of them might have come from a single parent that broke up during a close passage by the Sun. But while a single huge comet might break up into many pieces (Shoemaker–Levy 9 fragmented into 21 pieces before striking Jupiter), there have been over 2,000 of these sun-grazing comets detected by NASA's SOHO observatory alone (SOHO is the Solar and Heliospheric Observatory); it's possible that a single comet could be responsible for a number of progeny in similar orbits, but it seems unlikely that this one comet could be responsible for every single sun-grazing comet seen since.

These sun-grazing comets are intriguing objects – some pass so close to the Sun that they are actually inside the Sun's atmosphere, and drag from the atmosphere can be so substantial as to change the comet's orbit appreciably. Even more interesting, some of these comets do more than simply pass close to the Sun – some of them can evaporate completely during their passage; some even plunge into the Sun. Over time, any individual sun-grazing comet will eventually perish, breaking up,

Halley's Comet, photographed as part of the International Halley's Watch, 1986.

Launch of the SOHO spacecraft from the American launch facility at Cape Canaveral, Florida, 2 December 1995.

evaporating or falling into the Sun. But their ranks will be replenished – comets' orbits are fairly easily disrupted by the gravity of giant planets such as Jupiter, by the jets of gas they emit and (for the sun-grazers) by the atmosphere of the Sun during their closest approach. These changes sometimes push the comet's orbit closer to the Sun. In addition, there is a near-infinite reservoir of comets in the Kuiper Belt and Oort Cloud that can be perturbed by passing stars. These comets can fall into the inner Solar System from any angle and in virtually any orbit; just by chance some of these comets will be in sun-grazing orbits.

Despite the name, Aristotle did not discover this comet; he was only a child when it appeared in the sky. But even as a child, he may have been inspired by it – Seargent notes that, unlike his father (a physician), Aristotle turned his attention to nature and philosophy and exerted a profound influence over Western (and, later, Islamic) science that lasted for more than a millennium. But regardless of whether this comet is what inspired his interest in science, it certainly captured his attention – in his writing on comets, this was the one he considered to be the most impressive, the one he considered the greatest of them all.

44 BCE (Caesar's Comet)

In 44 BCE the Roman Empire was on the brink of civil war. Two years after Julius Caesar's assassination, various factions were vying for control of the empire. The next few years saw Antony and Octavian first join forces to defeat the armies of Brutus and Cassius, and then Octavian fought the forces of Antony and Cleopatra to win his empire. Two years into these civil wars, the Roman senate voted to deify Caesar, partly as a way of shoring up the support of the Roman population. During games held to honour Caesar's deification, a blazing comet appeared in the sky for a week. This appearance was seen as proof of Caesar's divinity, ultimately helping to cement Octavian's claim as Caesar's successor. The Roman poet Ovid commemorated the comet in *Metamorphosis*:

Then Jupiter, the Father, spoke . . . 'Take up Caesar's spirit from his murdered corpse, and change it into a star, so that the deified Julius may always look down from his high temple on our Capitol and forum.' He had barely finished, when gentle Venus stood in the midst of the Senate, seen by no one, and took up the newly freed spirit of her Caesar from his body, and preventing it from vanishing into the air, carried it towards the glorious stars. As she carried it, she felt it glow and take fire, and loosed it from her breast: it climbed higher than the moon, and drawing behind it a fiery tail, shone as a star.

The comet was reported by any number of other writers and historians over the next few centuries; Chinese astronomers also reported a bright comet around this same time – if contemporary accounts are to be believed, it was one of the brightest comets ever seen, but this could also be a misunderstanding of contemporary accounts. Many of these observations include the position of the comet in the sky; from these positions, it seems likely that the comet's orbit had an eccentricity of 1, meaning that it dropped into the inner Solar System once and escaped, never to return. But there are some problems.

One of the problems is that the observations do not exactly agree with the known calendar of events: there is a discrepancy between the dates of the comet recorded by the Chinese and the time of the games accompanying Caesar's deification. The Chinese astronomers were very attentive; there was not much that they missed, and it seems unlikely that they would have made a mistake in this, but the dates they reported seeing a major comet do not seem to coincide with the dates of Caesar's games. In fact, some astronomers have even gone as far as to suggest that the comet never existed – that Octavian fabricated the story for political reasons. But there is another possibility as well: Seargent notes that an eruption of Mt Etna at about the time of the comet's passage might have obscured its incoming leg, leaving it visible in Rome only while it was outbound after its nearest approach to the Sun.

soho image of a sun-grazing comet, photographed with the lasco c2 camera on 6 July 2011.

After so many centuries, there is no way to be absolutely certain about what actually happened; it seems likely that there was a bright comet that appeared shortly after Caesar's death and that this comet helped convince the public that Caesar had become a god. But exactly when the comet appeared in the skies and whether or not it is the same comet reported by the Chinese astronomers might never be known for sure. To some extent, however, this is a moot point – the comet was used to help cement Octavian's claim to the empire and helped shape the next several centuries of Roman history.

Roman coin minted shortly after Julius Caesar's death, showing the comet associated with his deification.

Comet Kirch (1680)

In the early days of telescopic observations, astronomers spent a fair amount of time simply looking at the skies to see what was there. Telescopes revealed details that nobody had even considered – the fact that the Moon has mountains and craters, that the Sun was not a perfectly unblemished sphere, that Saturn has rings (and Jupiter, satellites) and that the Milky Way contained stars too numerous to count. So it is not surprising that the earliest telescopic observers would turn their instruments to the skies at random, just to see what was there.

On 14 November 1680 Gottfried Kirch was looking through his telescope at the Moon and Mars when he noticed a fuzzy patch that looked a little odd. Within two weeks the comet had brightened to become one of the most dramatic comets of the century. Observations showed that the comet was headed on a near-collision course with the Sun. Then it vanished, lost in the glare of the Sun.

A few weeks later, observers saw another dazzling comet in the sky, this one headed away from the Sun. At that time there was still no real consensus about how comets moved through space, and astronomers had no reason to assume that the 'new' comet was the same one they had seen so recently on the other side of the Sun. Not only that, but it seemed unlikely that an object zipping through space could execute so tight a turn around the Sun – not until English astronomer John Flamsteed analysed observational data the following spring was there reason to believe that these two comets were one and the same. But not everyone was willing to believe Flamsteed's results – Isaac Newton in particular wrote to Flamsteed, 'To make ye comets of November and December but one is to make that one paradoxical.' Closer examination of Flamsteed's data, however, convinced Newton that he had erred; within a few years, he had accepted that the two comets were indeed one and the same.

There are some questions as to whether or not this comet is part of the Kreuz family of sun-grazing comets. On the one hand, the orbit has some significant differences from the other Kreuz comets, and one would expect that, if it was spawned from the same parent comet as the rest of the Kreuz family, it would have a very similar orbit. On the other hand, Kreuz himself noted that its orbit almost intersects that of the Kreuz comets. It is certainly plausible to suppose that it might, indeed, have originated from the parent Kreuz sun-grazing comet and, over the course of time, been pulled into a different orbit by the gravity of a planet it passed by, been pushed into another orbit by jets of expelled gas, or both.

Although the naked-eye observers could make general observations, this comet was the first to be studied by trained observers using telescopes, and the telescopic observations revealed details that had never been suspected. One observer

Comet Kirch in a
contemporary English
painting by an
unknown artist.

saw what he thought were fragments of the nucleus breaking
off and rejoining; this is entirely possible, but he could also have
been observing jets of gas spewing out of the comet and simply
did not know what he was seeing.

No matter how we look at it, Kirch's comet was remarkable.
It was blazingly bright and beautiful in the skies – one of the
brightest comets of the seventeenth century. It was the first

Eigentlicher abriß deß Schröcklichen Comessterns, welcher sich
den 26 December deß 1680 Jahrs von neuen widerümb sehen lassen, nach
dem er in die dritte worchen unter der Sonnen strahlen verborgen ge-
wesen Sein Lauf ist ziemlich geschwind und scheinet disem nach über den Mond
zu stehen der schweiff so viel man wegen der Helle des Monds abnehmen kön-
nen erstrecket sich auf die 70 grad. ferners bericht giebet das Tractätlein
von dem Urssprung des Cometen.
Zue finden bei Jacob Koppmeir buchtrucker in Aug:

comet to be discovered by telescope and to be telescopically
examined by trained scientists, setting the stage for centuries of
subsequent telescopic examinations of comets. It might well be
related to an entire family of comets, showing how much a
comet's orbit can be changed over its lifetime. And in his examination of Flamsteed's data (which, incidentally, was not entirely
with Flamsteed's permission), Newton came to realize that
comets travelled through space according to the same laws of

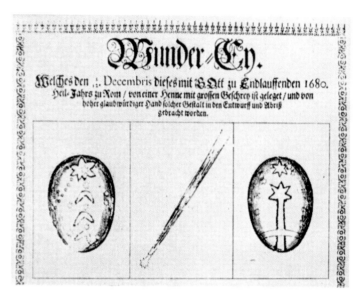

German woodcut showing the egg laid in Rome after the 1680 comet. There were many reports of such 'comet eggs' in Europe in these years; eggs with markings that were interpreted as representing or as being influenced by celestial events. In addition to Germany, such eggs were seen in Italy, France, Poland and elsewhere, sometimes accompanied by reports of unusual effort or cackling on the part of the hen.

motion as did the planets. Thus, Kirch's comet helped further Newton's understanding of the laws of gravity and planetary motion, with all of the profound implications that would ensue.

Hale–Bopp (1995)

On 26 March 1997, the San Diego police found 39 bodies in a mansion in the gated community of Rancho Santa Fe. The bodies were those of a local religious cult called Heaven's Gate, and they had committed mass suicide in order to reach heaven – they believed that an alien spaceship was going to transport their souls to a higher level of existence. They felt that this spaceship was accompanying a comet that had been in the news lately – comet Hale–Bopp.

Hale–Bopp was hardly a new comet when the Heaven's Gate members died; it had been discovered nearly two years earlier, in July 1995 by the amateur astronomer Alan Hale and novice Thomas Bopp. Astronomers quickly realized from its motion that it was still distant, which promised a brilliant appearance when it entered the inner Solar System and sprouted a tail. When it finally swung by the Sun, some astronomers estimated it to be hundreds of times brighter than the previous decade's Halley's Comet.

Although Hale–Bopp made a decided impact on society, its scientific impact was perhaps far greater. In particular, scientists studying the light from the comet and its tails discovered intriguing mixtures of gases that hinted at unexpected origins. From the amount of carbon monoxide the comet gave off (the same amount that would be emitted by over five billion cars daily), and the amount of water detected, astronomers became convinced that Hale–Bopp originated much closer to the Sun than the Oort Cloud – in fact, it was most likely born in the vicinity of Neptune and possibly even closer to the Sun. This, in turn, suggested to astronomers that at least some of the Oort Cloud comets were formed far closer to the Sun and were cast into the outer reaches of the Solar System by near encounters with massive planets – most likely Jupiter.

Although a surprise finding with regards to comets, it accorded with other speculations about the role of Jupiter and the gas giants in our Solar System; many astronomers now feel that Jupiter may have played a crucial role in keeping the Earth safe for life by clearing out the inner Solar System of bodies that might otherwise have exterminated life on Earth many times over, using its massive gravity to fling them into the outer reaches of the Solar System, where they are only an intermittent threat to life on Earth. When we consider that a rock only 10 km or so in diameter was enough to sound the death knell for the dinosaurs – along with about three-quarters of life on Earth – it should be clear that, absent such a shield, life would indeed have a difficult time surviving and becoming increasingly complex across billions of years. In any event, astronomers' studies of the chemistry of Comet Hale–Bopp were important in helping them to realize not only where comets were born, but how they got to be where they currently reside – and the role of giant planets in helping to maintain Earth as a habitable abode.

Comet Hale–Bopp had a decided impact on our society – people quite literally died because of what they thought it represented. But however great its cultural impact, by helping us to

Comet Hale–Bopp,
seen as a photographic
negative.

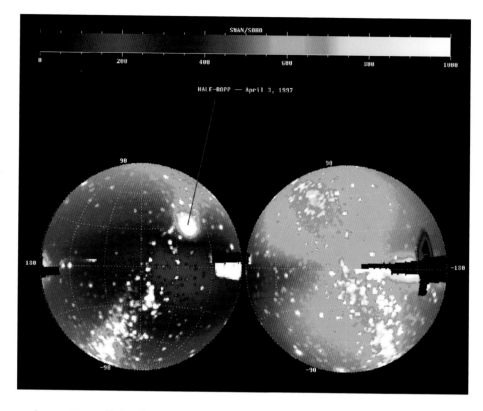

understand our Solar System better, its scientific impact was greater. As of this writing, Comet Hale–Bopp is well on its way into the outermost Solar System. We should see it again in the year 4380 (give or take a few decades).

McNaught (2007)

One of the most breathtaking comets in recent history was predicted to be rather a dud when it was first seen. Discovered on 7 August 2006 by the Australian astronomer Robert McNaught, the initial orbit calculations suggested that the comet would make it no closer to the Sun than Mars; even when later calculations showed a much closer approach inside the orbit of Mercury, comet-watchers calculated that it would be far too faint to excite much interest. As much as a few months later, in spite

False-colour SOHO image of Comet Hale–Bopp, showing the large extent of the hydrogen envelope of the comet's coma. The hydrogen comes from the evaporation of water ice and its subsequent dissociation by the solar wind.

of continued observations that suggested the comet might turn out to be brighter than predicted, it still looked to be too dim to excite much in the way of attention from the public. It turned out that the astronomers were wrong.

Looking at its orbit, astronomers concluded that Comet McNaught was making its first trip to the inner Solar System from the Oort Cloud (among the reasons for predicting a dim appearance was the observation that first-time visitors from the Oort Cloud have tended to make poor showings). But as it turned out, none of the predictions – whether based on experience, observations or calculations – was correct. McNaught continued to brighten more quickly than expected and, as it got closer to the Sun, it began to expel gas and dust at an increasing rate. By the time it reached the inner Solar System, it was hugely bright, with a magnificent tail that swept from horizon towards the zenith, and that arched dramatically, seemingly filling half the sky with bands of glowing dust laced with darker lanes where the dust was thinner. Any one of the bands would have made an impressive comet tail in and of itself; that the sky was filled with them made for a display that was nothing less than spectacular.

Unfortunately, Comet McNaught was visible to just a small fraction of the world's population because, due to its orbit, it could be seen only from the southern hemisphere when it was at its most spectacular. While Australians had a wonderful view of the comet, to North Americans and Europeans, only a part of the tail could be seen (although even this was eye-catching) once the comet dropped below the plane of the ecliptic and made its closest approach to the Sun. Within a few more months, the comet had left the inner Solar System on its return voyage to the Oort Cloud; it should reappear in our skies in about 90,000 years.

Unlike the other comets described in this chapter, Comet McNaught did not blaze new trails in science, it did not have a massive impact on society, and it failed even to have a huge cultural impact. What it did show us, though, was how far off the best scientific predictions could be – that what at first seemed

Comet McNaught
was one of the most
dramatic comets in
history.

to be a run-of-the-mill comet could end up as one of the most magnificent comets of the century.

One lesson from all of this is that comets have had a profound impact on humanity and the more spectacular the comet, the greater the potential impact. If, as Seargent suggests, Aristotle was inspired by the comet he saw as a child, then the comet named for him might have had a millennia-long impact on Western approaches to science. Caesar's comet might well have changed a millennium of European history by helping to cement Octavian's place in the Roman hierarchy, while both Comet Halley and Comet Kirch were instrumental in not only establishing telescopic observations of the cosmos, but inspiring

False-colour image of Comet McNaught captured by the SOHO LASCO C3 camera during the comet's approach to the Sun.

Newton's work in describing the universe in mathematical and mechanistic terms. Yet even a comet that serves 'merely' to capture the public's attention and causes them to lift their eyes towards the skies has still had an impact on our society – if only to remind the public of the wonders that lie in the heavens and that we can understand them so well.

7 Bringers of Life and of Doom

In addition to their scientific and cultural impact, comets have had a very literal impact on Earth (not to mention the rest of the Solar System). Comets are made primarily of water, and there are many planetary astronomers who suspect that comets brought water to the ancient Earth – that our oceans of today arrived with comets over four billion years ago during the Earth's infancy. Without comets, the water that is the basis for all life on Earth might not exist. Some scientists go even further: they have discovered that deep space holds complex molecules, including some of the amino acids and other compounds that form the basis of life. But comets can also bring death. We saw this in 1994 when Comet Shoemaker–Levy 9 slammed into Jupiter, creating Earth-sized bruises in the clouds shrouding the planet. Scientists have found evidence of cometary impacts on Earth as well – it is entirely possible that comets have caused extinctions on our planet in the past, and they might do so again in the future. Their role as bringers of doom was also feared during the apparition of Halley's Comet in 1910, when the Earth passed through the comet's tail, arousing a tremendous concern among the public.

Bringing life

In his Mars series of books, the award-winning science fiction author Kim Stanley Robinson postulates slamming comets into Mars to bring water to the planet as humans work to make it habitable.[1] The idea of harvesting interplanetary water is much older – Isaac Asimov's classic novella *The Martian Way*, of 1952, had Martian colonists travelling to Saturn to bring back a cubic mile of ice from the ring system to be used for the inhabitants' needs.[2] Casting comets at planets for terraforming, while not so common as to be a science fiction cliché, is an accepted plot point, readily understood by readers. Other science fiction stories have used similar plot points.

Comets have indeed been linked to the creation and sustaining of life on Earth. Comets are primarily ice, and water is essential to life as we know it – principally because it will dissolve virtually anything. Even gold, one of the least chemically reactive metals known, dissolves ever so slightly into water. Water does a much better job than other common solvents of dissolving the metals, minerals and more complex molecules that are vital to our existence. For such a simple molecule – water is only an oxygen atom joined to two hydrogens – it displays remarkably complex behaviour, both physically and chemically. We are more than half water by weight. Water carries nutrients from our digestive tract to our cells, and it carries waste from our cells to our kidneys; it carries the enzymes and hormones that keep our bodies working; it forms the bulk of our blood and our lymph; and our cells are filled with water. That is not to say that water is the only solvent that will support life – some have speculated that life might have evolved on Saturn's moon Titan, using hydrocarbon such as ethane as its solvent, but until we find (or fail to find) actual creatures on Titan, this is just speculation. For now, the only life we have found requires water, and there is water in abundance in the Solar System, including in the hundreds of billions (maybe trillions) of comets that fill the Kuiper Belt and the Oort Cloud.

Water

Talking about water is a good place to start. The universe seems to be full of water, but it has not always been that way; when the universe first formed, it contained nothing save hydrogen, helium and traces of lithium. The hydrogen itself came in two different

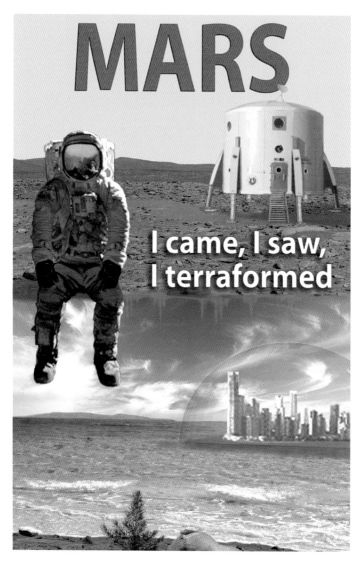

Artist's conception of what a terraformed Mars might look like; it is quite possible that the water would come from cometary impacts.

'flavours': the majority of hydrogen has only a single proton and a single electron, but a tiny percentage (about one-thousandth of 1 per cent, or one atom for every 100,000 atoms of hydrogen) contains both a proton and a neutron. This rare species of hydrogen is called deuterium, and it is a fragile atom that breaks apart when exposed to high levels of radiation, such as those that exist in the interiors of stars. For this reason, it can be used as a very sensitive tracer of where hydrogen (and any molecules such as water that contain hydrogen) comes from and whether or not it has ever seen the inside of a star.

When the Earth first formed, it likely contained water and other volatile compounds; the dust clouds from which it coalesced almost certainly contained water, and this would have been incorporated into the planet.[3] The early Earth likely contained oceans – the oldest rocks found show clear evidence of having been deposited underwater. The question is whether or not these early oceans have remained with the Earth until the present. Chances are that the water likely has not – and without water, no oceans. The impacts of a period in time called the Late Heavy Bombardment would have evaporated large amounts of water, some of which would have been broken apart into hydrogen and oxygen by the Sun's radiation; the hydrogen would have risen to the top of the atmosphere and eventually escaped into space. But the real ocean-killer was likely the formation of the Moon – there is good evidence that an object the size of Mars slammed into the Earth a few hundred million years after its formation, gouging out enough material to form the Moon. This impact very likely added enough energy to the Earth to melt the entire planet and to boil the oceans completely. Some of this water would probably have condensed again, but much would have been lost into space; replacing it took many millions (likely tens of millions) of years.

Earth contains somewhere between 1.3 and 1.4 billion km^3 of water. A large comet has a volume in the vicinity of 1,000 km^3 – filling the oceans would have required at least a few million comets' worth of ice. Considering that there has not been a major comet impact of anywhere near this size on Earth in recorded history, if the oceans were filled with water from comets, then there are a number of possible explanations: cometary impacts were far more frequent in the distant past; comets were much larger; the Earth was not stripped completely of water during the era of bombardments; or there was a combination of all of these. In fact, by looking at the isotopes found in water (isotopes are atoms with a specific number of protons and neutrons – stable carbon-12, for example, or radioactive carbon-14), scientists suspect that the water we have on Earth today is a mixture of our primordial water (what the Earth had when it first formed) and water brought by comets. The question is how they have come to this conclusion, and to answer that question, we have to understand more about where the atoms that make up the water – in particular, the deuterium (heavy hydrogen) atoms – come from.[4]

Stars get their energy from hydrogen – hydrogen fusion releases huge amounts of energy, powering the stars, and when hydrogen fuses, it forms helium. Helium, in turn, can fuse to form carbon, and so forth. These fusion reactions (called nucleosynthesis) are responsible for the presence of every atom in the universe heavier than lithium, including oxygen. Many stars, reaching the ends of their lives, expel large amounts of matter into the universe; this includes many of the heavier elements synthesized during the stars' lives. Interestingly, normal stellar fusion only produces atoms up to iron; anything heavier than iron takes much more energy and is only produced during the massive supernova explosions that mark the death of the largest stars.

Once these heavier elements have been flung into space by supernovae or wafted into space on stellar winds, they begin to collect to form molecules. Simple molecules are easier to form, and there are far more of the lighter elements than there are of the heavier. These two observations explain why, for example, water (composed of two hydrogen atoms and a single oxygen) is far more common than, say, lead oxide (lead is the heaviest non-radioactive atom), not to mention why methane (with a single carbon atom and four hydrogen) is more common than ovaline (a complex molecule with 32 carbon atoms and 14 hydrogen).

So water, a simple molecule that is made of light atoms, should be fairly common in the universe. And, indeed, when astronomers and planetary scientists look outwards, that is exactly what they find; water is found in relative abundance

All elements in the universe heavier than helium are formed by fusion inside of stars, as shown in this NASA graphic.

throughout the observable universe, usually in the form of ice, since the average temperature of the universe is barely above absolute zero.

The water found in the universe – including on Earth – includes a small percentage in which the hydrogen molecules are deuterium; these molecules are slightly heavier than typical water molecules and are what is known as 'heavy water'. In the universe in general, the fraction of water molecules made of heavy water should be about the same as the fraction of hydrogen atoms that are deuterium – about one in a hundred thousand or so (the actual fraction is less than this because some water molecules have had both hydrogen atoms replaced with deuterium). Since deuterium is destroyed inside a star, water made of hydrogen that has spent time inside a star will have even less deuterium than that in the universe in general. Every bit of water is a mixture of water formed of hydrogen atoms that have – and that have not – seen the inside of a star; each batch of ice will have a distinctive isotopic make-up reflecting the origin of its hydrogen atoms. By comparing the amount of deuterium in samples of water or ice, scientists can tell (roughly speaking) if they originated in the same place.

Astronomers have had the chance to study a few comets up close and to gather detailed spectroscopic information from some of them. Comparing the deuterium concentrations of these comets to that of the Earth's oceans might be able to tell us where the water in our oceans originated – if they are the same as those in comets, then the water on Earth likely came either from comets or from the same reservoir of water that went to form the comets; if the ratios are different, then the water had different origins.

In 1986 scientists got their first close look at a comet when they sent a flotilla of spacecraft to rendezvous with Halley's Comet. In future years not only were they able to visit more comets, but they could study them with ever-greater clarity from increasingly powerful telescopes and instruments, both on Earth and in orbit. And what they have found is intriguing; some comets have deuterium abundances that are similar to the universe as a whole (meaning that they are very likely formed from hydrogen that has never seen the inside of a star), some are similar to what is found in the Earth's oceans, and some resemble neither of these. With regards to the comets, this suggests that the water in comets is derived from a variety of sources, and that some of the hydrogen might date back to the formation of the universe but that some was likely processed by the stars before becoming locked into the comets' nuclei. But it also suggests that Earth's water is a mixture derived from comets as well as residual water that somehow survived the formation of the Moon.

Other molecules

Water is a start, but there is a lot more to forming life than just water. In addition to water, our bodies make extensive use of sodium, potassium, iron, carbon, nitrogen, magnesium, chlorine, phosphorus, sulphur and many more elements, as well as lesser use of ones such as selenium, cobalt, manganese, copper, lithium and a host of others. Of course, it is not enough simply to have these atoms in our bodies; they are incorporated into molecules – iron, for example, is part of the haemoglobin molecule that carries oxygen to our cells; our DNA is composed of phosphorus, sulphur, carbon, hydrogen and oxygen that are all arranged neatly into the double helix that carries our genetic information. The DNA tells our cells to link amino acids together in a particular order and pattern to form the proteins that make life possible.

The elements we need are all found on Earth as well as in the dust of which comets are (in part) comprised – finding iron, for example, in cometary dust is hardly a surprise. Simple molecules are also somewhat common in the universe – carbon monoxide (CO), methane (CH_4), ammonia (NH_3) and others fill the atmospheres of the gas giant planets and likely filled the atmosphere of the early Earth, in addition to making up many of the ices of which the comets are composed. What is a surprise is finding much more complex molecules in the cosmos, including a family of molecules called polycyclic aromatic hydrocarbons (abbreviated PAH) and amino acids. In fact, NASA's Stardust mission found traces of the amino acid glycine in Comet Wild 2, returning its

Amino Acids Produced by UV Ice Photolysis

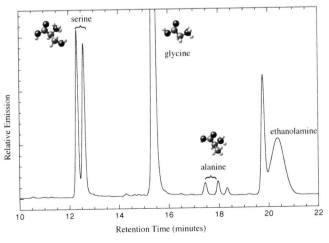

Researchers have found evidence that amino acids can be created in space and might have been delivered to Earth by comets and meteorites.

samples to Earth in 2009. But it is worth asking how such complex molecules could form in the icy depths of space.

Although chemical reactions proceed most quickly at relatively high temperatures – especially if the constituents are dissolved in a solvent such as water – they can take place even at the low temperatures of deep space and even in the absence of a solvent. So picture a comet deep in the outer Solar System: it will contain large amounts of the light elements – hydrogen, oxygen, carbon, phosphorus, sulphur and so forth – and, although these elements will be at incredibly low temperatures and not floating in a solvent, given enough time, they will link up. Aided

A number of different amino acids have been found inside meteorites associated with comets.

Comet dust particles captured and returned to Earth by NASA's stardust mission.

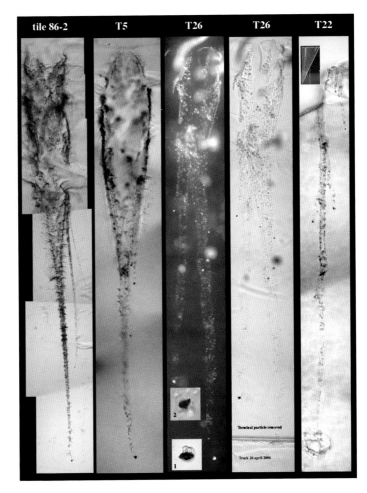

and abetted by ultraviolet radiation from the Sun, even ice chilled to the temperatures that exist between the stars will support fairly complex chemical reactions. Scientists have shown that molecules that are the precursors to life can form on comets, and that these molecules can survive the fiery plunge through an atmosphere, along with the (literally) earth-shattering crash at the end of the journey. In short, at least some of the complex chemicals that form the basis of life were very likely brought to the early Earth by comets and could well have shortened the time needed for life to evolve on our planet.

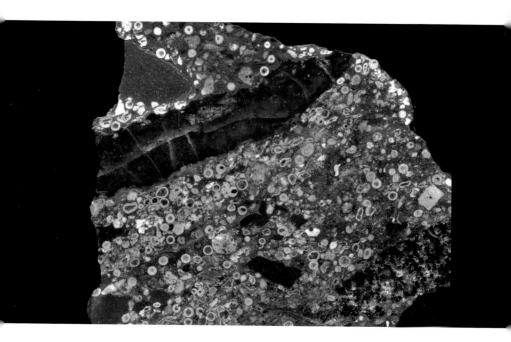

The earliest Earth almost certainly had water, all of the chemical elements (including the ones needed by living organisms) and even a number of the chemicals of life. In fact, in 1952 the chemists Stanley Miller (1930–2007) and Harold Urey (1893–1981) conducted a classic experiment in which they showed that the chemicals and environmental conditions of the early Earth could have formed every amino acid we have today, and many more.[5] What Miller and Urey did not know was that the early Earth was pummelled mercilessly by the Late Heavy Bombardment (which lasted for a few hundred million years), and any life that might have formed before that time was almost surely exterminated. Many have speculated that whatever water was part of the Earth when it formed would have been destroyed by the heat of these collisions, not to mention whatever complex molecules might have taken shape. According to some thinking, this water and these molecules were replaced by comets colliding with the Earth. Thus comets might well have provided many of the ingredients essential to the genesis of life – and to our being here at all.

The small spherules in this rock are debris from a major impact that was blasted briefly into space and fell back to Earth more than 2.5 billion years ago. They were discovered in 2004 by Bruce Simonson of Oberlin College; in 2011 Purdue University scientists Jay Melosh and Brandon Johnson interpreted the discovery as representative of an impact.

Bringing doom

The science fiction authors Larry Niven and Jerry Pournelle wrote one of the best comet disaster books of all time, *Lucifer's Hammer*, published in 1977.[6] In this book, a comet is found and predicted to pass close to the Earth; astronomers try to predict its exact trajectory, but jets emitted by vaporizing volatile compounds keep pushing the comet off course. As the comet approaches the Earth, it breaks into myriad smaller bodies, striking the Earth in multiple locations at sea and on land. Land strikes are bad enough, causing earthquakes, setting fires and pulverizing whatever happens to be beneath or nearby; sea strikes are even worse, raising tsunamis and huge clouds of steam. With all of the steam and dust in the atmosphere, global temperatures start to drop, threatening a cometary equivalent of nuclear winter. As if the natural disaster were not enough, Russia and China launch nuclear missiles at each other, and governments worldwide fall, giving rise to smaller-scale conflicts, warlords and limited fiefdoms. *Lucifer's Hammer* was widely read and well liked by readers; a number of scientists also read it approvingly, noting that Niven and Pournelle did a pretty good job of getting the science right. In fact, in a few of their plot points, they were prescient – suggesting that a comet could break apart predated the breakup of Shoemaker–Levy 9 by nearly two decades – and they were also ahead of most scientists in speculating about impact-caused mass extinctions. *Lucifer's Hammer* is a work of fiction, of course, but reality might be far worse. In fact, Niven and Pournelle might have been optimistic; more recent work suggests that a comet strike of this size might be enough to wipe out humanity – and much of the life on Earth – entirely.

In the late 1970s, geologist Walter Alvarez (b. 1940) was examining rocks in the Italian Alps and puzzling over an odd layer of clay.[7] In the rocks beneath the clay layer were the bones of dinosaurs and a rich collection of other fossils; all of these vanished above the clay layer. The clay appeared to mark the boundary between the age of the dinosaurs and their extinction. Looking at the chemistry of the clay, Alvarez noticed an odd

concentration of a rare element – iridium – and speculated about what it might mean. With the help of his father, the Nobel laureate Luis Alvarez (1911–1988), and others, Alvarez was able to puzzle it out: iridium is more common in extraterrestrial objects than it is in the Earth's crust; the most plausible answer was that this clay was laced with debris from a cosmic collision just at the time of the dinosaurs' extinction. In all likelihood, this collision is what killed off the dinosaurs. Although impact craters were well known (the planetary geologist Gene Shoemaker had recognized that extraterrestrial bodies could create impact craters in his doctoral dissertation of 1960), nobody had linked any such impact with the dinosaurs' demise. In fact, until Alvarez's discovery, nobody had linked cosmic impacts with *any* extinction, let alone the most famous of them all. Prior to his

A Wyoming rock on display at the San Diego Natural History Museum. It has a layer of clay (light-coloured stripe) that contains over 1,000 times as much iridium as the layers on either side.

work, geologists eschewed the idea that a single catastrophic event might have such an overwhelming effect; a single scientific paper reintroduced the idea that geologic change might be invisibly incremental most of the time, and catastrophically quick every now and again.

The Alvarezes' work was compelling and it convinced scientists that extraterrestrial impacts could alter life on Earth, but it was something that had happened far in the past. It was easy to speculate that the Earth – and the Solar System – had changed and that impacts were no longer an important consideration. And, while the object postulated by Walter and Luis Alvarez was stony (an asteroid), it was easy to apply this same logic to icy objects such as comets. As it turned out, less than two decades later, astronomers – indeed, anyone with access to cable television or the Internet – saw a comet (co-discovered by planetary scientist Gene Shoemaker) slam into Jupiter, throwing material far into space and leaving welts the size of the Earth that were clearly visible from anywhere in the Solar System. The

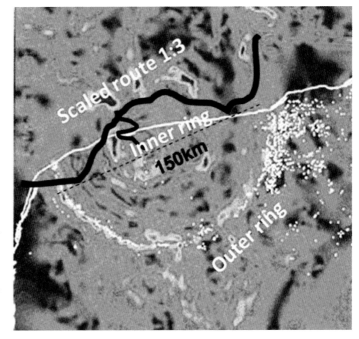

Gravitational anomaly map of the Chicxulub impact crater on the Yucatán Peninsula. This impact is credited with having led to the extinction of the dinosaurs.

comet, discovered just two years earlier, had broken into 21 pieces while orbiting Jupiter; on 16 July 1994 the first of these fragments struck Jupiter's atmosphere, and the other twenty pieces followed during the next few days. Some fragments, photographed by the Hubble Space Telescope, raised plumes thousands of kilometres high; others left dark blotches larger than the Earth. Nobody who witnessed this series of impacts could doubt that comets are fully capable of causing mass extinctions, nor that such events can still take place. In the words of American astrophysicist Neil deGrasse Tyson, we dare not forget that we live in a cosmic shooting gallery.

If we look around the Solar System, we see ample proof of Tyson's comment – the Moon, Mars, Venus, Mercury and all the major moons of the Solar System are pocked with craters from impacts – most of these impacts seem to have happened billions of years ago, during the Solar System's earliest days, but planetary scientists saw a bright flash signifying a large impact on the Moon as recently as 2013. Even on our home planet, we see evidence of major impacts: Earth has no fewer than 44 confirmed impact features larger than 20 km (12 miles) in diameter, with the largest a whopping 300 km (186 miles) across. The oldest of these craters is nearly half the age of our planet (slightly over two billion years old), but most are relatively young because the majority of the rock formations on Earth are relatively young – not many surface features survive more than a few tens of millions of years' of erosion and plate tectonics, while most marine features are either covered with sediments or drawn down into the Earth as a result of tectonic activity.

Interestingly, some of these impact features are close to the same age and, in some cases, even show signs of being aligned; this suggests that they were made by objects that broke apart before striking the Earth. (The 213-million-year-old Manicouagan impact crater in Canada seems to be linked to others in France, Manitoba, North Dakota and Ukraine.) In some cases, geologists have to reconstruct where the continents were at the time of the impacts, but when they do so, they find a striking pattern that looks very much like the series of welts left in the clouds of

Jupiter by Shoemaker–Levy 9. At the very least this tells us that it is not unheard of for celestial objects to break up, causing multiple impacts from a single parent object. In addition, it points out to us a danger in trying to divert any asteroid or comet on a collision path with Earth: we want to avoid taking actions that could break up what we're trying to divert lest we turn a single catastrophic impact into multiple huge impacts scattered across the globe.

Comets – or any celestial object – can do so much damage because they carry an enormous amount of energy with them. Consider a comet made of ice with a diameter of 20 km. The comet will have a volume of about 4,200 km^3 and will weigh about 4,200 billion tonnes. Moving at 20 km/sec, the comet will have the energy of more than 100 million hydrogen bombs and will lift dust high into the atmosphere, blocking the Sun for years or decades. The ensuing winter, in conjunction with other environmental impacts, would be fatal to much of the life on Earth – the loss of sunlight would cause plants to wither and die, and everything that fed on plants (or on herbivores) would die as well. But this is only the start.

To get a little into the physics of these collisions, the amount of damage is going to be proportional to the mass of the object (comet or asteroid) that strikes the planet. All other things being equal, the mass scales as the cube of the size of an object. (Cubing a number means you multiply it by itself three times – for example, 2 × 2 × 2 = 8, so two cubed is equal to eight.) If you double the size of a comet – say from 5 to 10 km – you increase the mass by a factor of eight, and you increase the destruction it can cause by the same amount. The amount of energy increases even more rapidly as the speed increases, but the speed at which objects strike the Earth does not seem to vary nearly as much as the size of the objects.

The heat of the comet's passage through the atmosphere and its impact into the planet will cause massive chemical reactions to occur and will throw material into the atmosphere. At the very least, the heat will cause nitrogen and oxygen to combine to form nitrous oxides; these will cloud the atmosphere and reduce sunlight reaching the ground. Some of the nitrous oxides will

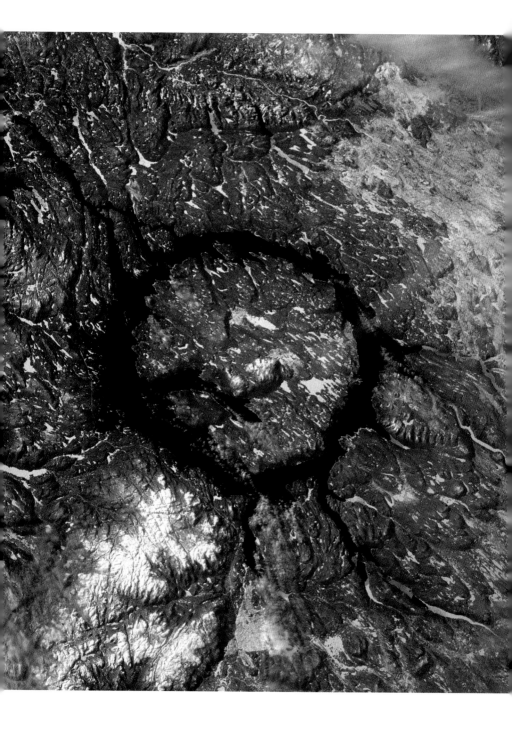

More than 200 million years ago a huge impact formed a 100-km (60-mile) diameter crater on the Canadian Shield that has now filled with water to form the Manicouagan Reservoir.

dissolve into water vapour in the atmosphere to form nitric acid; when the water vapour condenses to form rain, it will be dangerously acidic – more so than the acid rain that so badly damaged the American Adirondack Mountains and mountain lakes in the 1980s and 1990s.

If the acid rain falls on land, it will speed up erosion by dissolving many of the rocks it falls on; the material dissolved will ultimately end up in the oceans, changing their chemistry somewhat. The acid rain that falls on vegetation will kill most of the plants, upsetting the food chain. And everything that falls on the ocean will wreak havoc with the surface organisms, again wreaking havoc on the oceanic food chain.

At the same time, the superheated air will rise into the atmosphere, reaching the stratosphere and higher. Here, the nitrous oxides will cause still more chemical reactions – in this case, destroying the ozone molecules with which they come in contact. Within weeks, most – possibly all – of the Earth's ozone layer will cease to exist, at least until it re-forms decades or centuries in the future. Meanwhile, any plants and animals sensitive to ultraviolet (UV) radiation will be harmed or killed.

At some point, the ozone will re-establish itself, the dust will settle, and the air will clear. Eventually the organisms that were not killed will reproduce and will start to fill the world again. Ultimately, too, the clouds of nitrous oxides will clear, and temperatures will begin to rise back to normal. And actually, they might continue to rise – all of the water vapour left in the atmosphere will act as a greenhouse gas, and it is entirely possible that carbon dioxide levels will rise as well, released by burning plants in the immediate aftermath of an impact on land. It will be decades or centuries before temperatures return to normal.

So this is what a comet will bring to Earth if – when – one

The largest of the 21 fragments of Comet Shoemaker–Levy 9, taken shortly before the leading fragment struck Jupiter.

SL-9
INCLINATION : 94.233°
IMPACT SPEED : 60 KM/S
FRAGMENTS : 21 PRIMARY

strikes again. Dust, acid rain, the loss of our ultraviolet protection and the death of countless plants and animals. An 'impact
winter' followed by a greenhouse planet will test all but the har-
diest of organisms. Life will first freeze and will then roast, and
only the rugged will survive. Given the stakes – the potential
extinction of humanity and much of life on Earth (a possible
impact at the end of the Permian period, about 252 million
years ago, destroyed as much as 95 per cent of all living species),
it makes sense to wonder if there might be any way to avert
catastrophe.

In the movies, diverting asteroids (the same techniques can
be applied to comets) is fairly simple – blow them out of the sky
with nuclear weapons or use nuclear weapons to blast them into
a new orbit. Appealing as this might be, the chances of success
are incredibly low, while the chances to make things worse are
fairly significant.

When a nuclear weapon detonates, it forms a fireball; inside
this fireball temperatures are high enough to vaporize virtually
any substance known. So it's reasonable to think that a suffi-
ciently large nuclear weapon might simply vaporize a comet on
a collision course with Earth. But the largest nuclear weapon yet
detonated – the Tsar Bomba, with a yield of over 50 megatonnes
– had a fireball diameter of about 8 km; were it set off inside a

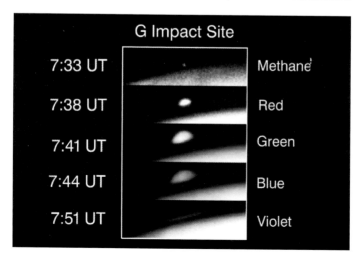

The rise of a plume
of superheated gas
from Jupiter following
the impact of the
G fragment just after
8 pm (GMT) on 16 July
1994. Photographed
with the Hubble
Space Telescope's
Wide Field Planetary
Camera.

comet, the fireball would likely have been somewhat smaller. First, such a device would, at most, be unable to vaporize anything larger than this fireball, and even that would require drilling a hole to the centre of the comet to set off the device internally. Any larger object would be partially vaporized and broken apart into tens or hundreds of large fragments and innumerable smaller ones. Consider: a fireball with a diameter of 8 km will have a volume of close to 50 km^3, while a comet with a diameter of 10 km will have a volume of about 75 km^3. This means that even if everything within the fireball were turned to vapour, there would still be a third of the volume of the comet left to strike the Earth – and this is for a comet just barely larger than the size of the fireball, assuming that we can slap together a 50 megatonne bomb, and assuming that everything within a 25 km radius is completely vaporized! In reality, it is far more likely that over half of the comet would still be around to pummel the Earth.

Destroying a comet might not be possible and could make things worse – what about using a bomb to blast it off course? After all, a comet has an appointment with Earth and has to be in a precise place at the designated time; the Earth is only about 13,000 km across, and it moves through space at a speed of about 100,000 km/hour – we only need to delay (or speed up) a comet enough to have it cross Earth's orbit about seven or eight minutes later or earlier in order for it to miss the Earth, or shove it to the side by 6,000–7,000 km to turn a disaster into a near-miss. Here's the problem with that: what if we make a mistake and turn what would have been a near-miss into a direct hit? And an attempt to divert a comet explosively might still end up fragmenting it. No matter how we look at it, explosives are not the best way to try to save the Earth, except perhaps as a last resort.

On the other hand, we might not need anything so dramatic. Why try to blast a comet off-course when we can gently push or pull it instead? Given enough warning (a decade, say), it might be possible to put a low-thrust rocket engine on a comet to push it off-course, to speed (or delay) it in its orbit – whatever it takes to make it miss the Earth. Remember that we only need delay it (or speed it up) by a handful of minutes to make it miss our planet. Even a very small thrust applied over enough time can have a substantial influence on an orbit.

Yet another possibility is to send what Tyson calls a 'gravitational tractor' – a massive spacecraft that would be sent to rendezvous with the comet – nudged into place and maintained there with small thrusters. According to gravitational theory, the 'tractor' would exert a pull (albeit a small one) on the comet. By carefully positioning the tractor and through precise station-keeping, it can be used to pull the comet away from its rendezvous with Earth very slowly.

There are other options as well that have been suggested: positioning mirrors to create jets to push the comet off-course, colouring one side of the comet black to increase the absorption of sunlight and heat one side preferentially, and so forth. But these options seem less feasible than the ones mentioned. Incidentally, the same factors that made the fictional comet so deadly in *Lucifer's Hammer* can make it challenging to carry out any diversionary scheme. If a gas jet erupts on the opposite of the comet from a small rocket motor it can completely cancel out the rocket's efforts. Along the same lines, the random nature of these jets (we simply cannot predict when or where they will erupt) can also make it difficult to redirect a comet precisely. For this reason, if time permits, it makes sense to try to push it as far off-course as possible rather than going for a near-miss.

To sum up, a collision with even a relatively small comet can have deadly consequences for life on Earth; a relatively minor consequence would be the end of civilization as we know it, and extraterrestrial impacts have caused mass extinctions in which the majority of life died out. We see evidence of impacts on Earth and on virtually every object in the Solar System, and we know that these impacts are continuing to take place even today. It is inevitable that Earth will be struck again – the only question is when that will happen. Although explosives (including nuclear explosives) are not a good idea, since they might cause a comet to break up, there are other, gentler options to consider: mounting small rocket engines or using a gravitational tractor to push or pull a comet off course by the relatively minor amount needed to turn an impact into a near-miss, for example. On the other hand, given the inevitable uncertainties in predicting a comet's orbit – especially the unpredictable eruption of jets from gas pockets – we might need to aim for more than a near-miss in order to ensure our protection.

One final point: a large enough impact can wreak catastrophic damage on the Earth's ecosystem, but there is a limit to the amount of damage it can cause. It is entirely plausible to think that an impact could destroy our technological civilization; an impact of the magnitude of the one 65 million years ago (which contributed to one of the Earth's five mass extinctions) can even cause our extinction and that of many other organisms. But life has proven itself to be remarkably resilient over the ages – life has been present on Earth for nearly four billion years, and in all that time, it has never been totally extinguished; it would take more than a single comet to wipe out all life on Earth. To do that would take an impact by an object the size of the Moon – something large enough to vaporize the oceans and to melt the Earth's crust. To the best of our knowledge, there are no unaccounted bodies of that size in our Solar System.

Conclusion

Comets carry with them clues about how our Solar System was formed; they have helped to teach us how objects in the universe move; and we have learned about how they might have brought not only death and destruction, but life to Earth and possibly to other bodies in our Solar System and beyond. Comets might or might not be messengers from the gods, but they are messengers from distant places and from distant times, bringing us information about places and times that would otherwise be far beyond our grasp and our ken.

The study of comets has not only helped our scientists to elucidate the history of our planet and our Solar System, but the workings of our universe. Newton's laws of motion and his law of gravity were developed using, among other things, observations of comets, and they have been found to apply to our planet, our Solar System, our galaxy and the rest of the universe. When astronomers looked outwards and began measuring the speed of galactic rotations, it was Newton's laws of gravity and motion that forced them to acknowledge the presence of supermassive black holes in the core of virtually every galaxy we have seen, and these same laws of motion forced them to acknowledge that galaxies' motions can only be explained by the presence of vast amounts of unseen matter – dark matter – permeating space. In fact, the vast majority of matter in the universe is composed of material whose presence can only be deduced using the laws of motion expounded by a scientist of the Enlightenment. Would Newton (or someone else) have developed these same laws had no comets existed in our Solar System? Of course he would have, but the fact remains that high-quality cometary data existed in large part because of the beauty and the puzzle posed by these itinerant visitors, and these data were hugely helpful in teasing out the answer to the puzzle of the movement of cosmic bodies.

So we have learned a tremendous amount of science from the study of comets, but we have equally learned about ourselves in the way that we portray comets in our art, in literature and in the way that comets affect our religious thought. We find that we want to find order in the universe, and, in the absence of scientific and mathematical tools, we find this order by postulating that the objects we see in the skies must be messengers from one or more gods (depending on one's religion), or they must convey to us some degree of cosmic influence and messages from the stars. At first blush, religion (including cults) and pseudoscientific practices such as astrology seem as opposite to scientific inquiry as one might imagine. Yet both represent humanity's desire to understand our universe, to understand the mechanisms by which it operates and to tease out whatever patterns might emerge in order to help us to predict what the future might hold. Whether we hold the prime movers of our universe to be natural or supernatural, we are still trying to accomplish the same thing: to understand these movers in a way that makes sense to us and in a way that can be useful to someone, somewhere. Whether the end user is an astronomer trying to predict (and hopefully avert) a civilization-ending collision, an astrologer trying to predict future events, a shaman trying to predict the optimum time to plant crops (or to go to war) or a priest trying to understand the will of his (or her) god, our study of comets through the ages is an attempt to understand how our universe works and to use that knowledge for our own purposes, be they good or ill.

Finally, in art of all types – writing, photography, painting and more – comets can be used as they are, or they can be freighted with more significance. Here, the process of discovery continues, but the comets help us to understand ourselves as much as the world and universe around us. Their use in some contexts shows us that, even in a scientific age, we are still susceptible to the mystic; their use in the past shows us how we have grown as a species. We once saw the world as a frightening place filled with signs and portents and run by capricious deities and occult forces whose thoughts could only be understood by a select few priests and practitioners. We have come to realize that we reside in a universe that operates by physical laws that can be, to a large extent, understood by most of us. But whether a painting shows a comet as a sign from heaven or simply as a pretty object in the sky, it shows us how the artist – and by extension, how the society of the day – viewed not only comets, but the universe in which they appear, and how they viewed our ability to understand them. Art, ultimately – regardless of the subject – tells us about ourselves.

So perhaps this is the central theme of our foray into comets: we look at them to discover. We discover more about the universe in which we live, we discover our relationship with our universe, and we discover how we see ourselves as citizens of this universe. Comets are part of the key to understanding not only the universe, but ourselves as well. And this might well be as good a conclusion to draw as any.

REFERENCES

1 The Science of Comets

1 M. A. Barucci, H. Boehnhardt, D. P. Cruikshank and A. Morbidelli, eds, *The Solar System Beyond Neptune* (Tucson, AZ, 2008).
2 Fred L. Whipple, 'A Comet Model. 1. The Acceleration of Comet Encke', *Astrophysical Journal*, CXI (1950), pp. 375–94.
3 Michael J. S. Belton, 'Whipple's Comet Model', *Astrophysical Journal*, DXXV (1999), pp. 393–4.
4 M. C. Festou, H. U. Keller and H. A. Weaver, eds, *Comets II* (Tucson, AZ, 2004), pp. 3–16.
5 Professor Bradley Peterson, in an undergraduate class the author attended, 1990.

2 Studying Comets through the Ages

1 Sara Schechner Genuth, *Comets, Popular Culture, and the Birth of Modern Cosmology* (Princeton, NJ, 1997), pp. 17–19.
2 A. A. Barrett, 'Observations of Comets in Greek and Roman Sources before AD 410', *Journal of the Royal Astronomical Society of Canada*, LXXII/2 (1978), pp. 81–106.
3 Jane L. Jervis, *Cometary Theory in Fifteenth-century Europe* (Dordrecht, 1985).
4 'Libyan Desert Glass: Diamond-bearing Pebble Provides Evidence of Comet Striking Earth', Science-News.com, 8 October 2013, www.sci-news.com, accessed 8 March 2016.
5 Donald K. Yeomans, *Comets: A Chronological History of Observation, Science, Myth, and Folklore* (New York, 1991), pp. 51–68.
6 Auguste Comte, *The Positive Philosophy*, Book II, Chapter One (1842).
7 Michael J. S. Belton, 'Whipple's Comet Model', *Astrophysical Journal*, DXXV (1999), pp. 393–4.

8 Jan H. Oort, 'The Structure of the Cloud of Comets Surrounding the Solar System, and a Hypothesis Concerning its Origin', *Bulletin of the Astronomical Institutes of the Netherlands*, XI/48 (1950), pp. 91–110.

3 Visualizing Comets

1 Fernando Coimbra, *The Sky on the Rocks: Cometary Images in Rock Art*. Published in the *Proceedings of the Global Rock Art International Congress* (Piauí, Brazil, 2010), pp. 635–46. Fernando Coimbra, 'Astronomical Representations in Rock Art: Examples of the Cognitive and Spiritual Processes of Non-literate People', in *Proceedings of the Intellectual and Spiritual Expressions of Non-literate Societies*, ed. Emmaunel Anati, Luiz Oosterbeek and Federico Mailland (Florianopolis, Brazil, 2011), pp. 37–44.
2 Carl Sagan and Ann Druyan, *Comet* (New York, 1997), pp. 172–97; and Fernando Coimbra, 'The Astronomical Origins of the Swastika Motif', in *Proceedings of the Intellectual and Spiritual Expressions*, ed. Anati, Oosterbeek and Mailland, pp. 78–90.
3 Carl Sagan and Ann Druyan, *Comet* (New York, 1985), p. 184.
4 Roberta J. M. Olson, *Fire and Ice: A History of Comets in Art* (New York, 1985), pp. 26–49.

4 Comets and Religion

1 Jonathan Flanery, 'Unexpected Visitors: The Theory of the Influence of Comets', www.skyscript.co.uk/comet.html, accessed 11 April 2015.
2 Roberta J. M. Olson, *Fire and Ice: A History of Comets in Art* (New York, 1985), pp. viii–ix. Emphasis in original text.
3 Kevin Curran, *Fall of a Thousand Suns: How Near Misses and Impacts by Comets Affected the Religious Beliefs of our Ancestors* (Kindle edn, 2014).
4 John Headley Brooke, *Science and Religion: Some Historical Perspectives* (Cambridge, 1991), p. 460.

5 Comets in Literature and Popular Culture

1 Sara Schechner Genuth, 'Astronomical Imagery in a Passage of Homer', *Journal for the History of Astronomy*, XXIII (1992), pp. 293–8.
2 Sara Schechner, 'Astronomical Imagery in a Passage of Homer', *Journal of the History of Astronomy*, XXIII/4 (1992), pp. 293–8.
3 Torquato Tasso quoted ibid.
4 Camille Flammarion, *La Fin du Monde* (1894), available at http://archive.org.

5 H. G. Wells, *In the Days of the Comet*, available as a free electronic book through the Gutenberg Project, www.gutenberg.org/ebooks/ 3797, accessed 11 April 2015.

6 **History's Greatest Comets**

1 D. Justin Schove, *Chronology of Eclipses and Comets*, AD *1–1000* (Bury St Edmunds, 1984), p. 285.
2 David Seargent, *The Greatest Comets in History: Broom Stars and Celestial Scimitars* (New York, 2009).
3 Nigel Calder, *The Comet is Coming! The Feverish Legacy of Mr Halley* (London, 1980), pp. 24–6.
4 Seargent, *The Greatest Comets in History*, p. 192.

7 **Bringers of Life and Doom**

1 Kim Stanley Robinson, Mars Trilogy (*Red Mars*, *Green Mars* and *Blue Mars*) (New York, 1993, 1995 and 1997).
2 Isaac Asimov, *The Martian Way* (New York, 1981).
3 A. Morbidelli, et al., 'Source Regions and Timescales for the Delivery of Water to the Earth', *Meteoritics & Planetary Science*, xxxv (2000), pp. 1309–20.
4 Nicolas Dauphas, François Robert and Bernard Marty, 'The Late Asteroidal and Cometary Bombardment of Earth as Recorded in Water Deuterium to Protium Ratio', *Icarus*, cxlviii (2000), pp. 508–12.
5 Stanley L. Miller, 'A Production of Amino Acids under Possible Primitive Earth Conditions', *Science*, cxvii/3046 (1953), pp. 528–9.
6 Larry Niven and Jerry Pournelle, *Lucifer's Hammer* (New York, 1985).
7 Luis W. Alvarez et al., 'Extraterrestrial Cause for the Cretaceous-tertiary Extinction', *Science*, ccviii/4448 (1980), pp. 1095–108.

SELECT BIBLIOGRAPHY

Books

Eichner, David, *Comets! Visitors from Deep Space* (Cambridge, 2013)
Levy, David, *Comets: Creators and Destroyers* (New York, 1998)
Reynolds, Alastair, *Revelation Space* (New York, 2002)
Simmons, Dan, The Hyperion Cantos (*Hyperion, The Fall of Hyperion, Endymion* and *The Rise of Endymion*) (New York, 1990, 1995, 1996 and 1998)
Stoyan, Ronald, *Atlas of Great Comets* (Cambridge, 2015)
Verne, Jules, *Hector Servadac's Voyages and Adventures across the Solar System.*

Scientific papers

Hernández, Orlando, Sait Khurama and Gretta C. Alexander, 'Structural Modeling of the Vichada Impact Structure from Interpreted Ground Gravity and Magnetic Anomalies', *Boletín de Geología*, XXXIII/I (2011). Available online at www.scielo.org.co, accessed 17 May 2015
Science Magazine Special Issue: Catching a Comet, CCCXLVII/6220 (23 January 2015). Available online at http://science.sciencemag.org, accessed 7 March 2016. This is a special issue of scientific papers on the Rosetta mission to Comet 67P/Churyumov–Gerasimenko. These papers describe the mission itself (including the soft landing) and what has been learned as of that date from the mission.

Internet resources

Anasazi Photography, http://anasaziphoto.com/, accessed 25 April 2015
Astrobril, www.astrobril.nl, accessed 10 May 2015
Astrology and Religion Among the Greeks and Romans, http://sacred-texts.com, accessed 11 April 2015
Atlas Coelestris, www.atlascoelestis.com, accessed 10 May 2015

'Comets: Facts about the Dirty Snowballs of Space', Space,
     www.space.com, accessed 12 April 2015
'Comet Shoemaker–Levy 9 Collision with Jupiter', NASA,
     http://www2.jpl.nasa.gov/sl9/, accessed 17 May 2015
'Comets', Sky and Telescope, www.skyandtelescope.com, accessed
     12 April 2015
'Deep Impact Mission to Comet Tempel 1', NASA, www.nasa.gov,
     accessed 11 April 2015
'Giotto Mission to Halley's Comet', ESA, http://sci.esa.int/giotto/,
     accessed 11 April 2015
Gregg's Astronomy, www.greggsastronomy.com, accessed 25 April 2015
RockArt Blog, http://rockartblog.blogspot.com/, accessed 25 April 2015
'Rosetta Mission to Comet 67P/Churyumov–Gerasimenko', ESA,
     www.esa.int, accessed 11 April 2015
'Stardust Sample Return Mission to Comet Tempel 1', NASA,
     http://stardust.jpl.nasa.gov, accessed 11 April 2015

ACKNOWLEDGEMENTS

As is the case in any work – especially one that covers as much ground as does this – thanks go out to the many people who took time from their schedules to help me. While many offered their help, a few went above and beyond expectations and deserve special mention.

Michael Leaman of Reaktion Books was kind enough to offer me a chance to write about a fascinating topic, was generous with his time in helping me to develop this book and was patient as deadlines approached. Bob Diforio has ably represented me for several years, helping me at all stages of the writing process. I appreciate both his help and his counsel over the years. Professor Fernando Coimbra at the Quaternary and Prehistory Group – Geosciences Center of the University of Coimbra, Portugal has spent years studying prehistoric art, including a number of instances in which comets were depicted in art on rocks. Professor Coimbra was generous with his time, including sending me copies of many of his academic papers on the topic. Jonathan Flanery is a professional astrologer and is one of the few to write about the astrological significance of comets. Mr Flanery was kind enough to spend time helping me to understand how comets figure into the practice of astrology, as well as giving me permission to quote him at length. Detective Marcos Quinonas of the New York City Police Department is an expert on a variety of occult religions; he helped me to understand the significance of comets to religious cults and less mainstream religions. Dr Neil deGrasse Tyson generously spent time with me during an interview, discussing (among other things) the risks that Earth faces from impacts.

Mrs Lizzie Calder and her daughter Penny were very kind to let me use an image of the cover of *The Comet is Coming*, written by the late Nigel Calder. I thank my daughter Teresa Nguyen for her help with graphics, including the sketch of a comet in Chapter One. Randy Langstraat was kind enough to permit the use of his photos of petroglyphs featuring comets. Other work of his can be found on his website, Anasazi

Photography (http://anasaziphoto.com/). Gregg Ruppel graciously granted me permission to use his photos of Comet c2011 to help illustrate how scientific photography can become art. Yong Lin Tan was equally generous in permitting me to reproduce his artwork, much more of which can be found on his website (http://yongl.deviantart.com/).

Additional thanks go to Dr Richard Pogge, Professor Orlando Hernández Pardo, Ms Teri Smoot and the Star Shadows Remote Observatory, the Utica Comets hockey team and others who were kind enough to let me use their graphics to illustrate various concepts in this book. The European Southern Observatory and the European Space Agency were kind enough to let me use some of their photos, as was the National Optical Astronomy Observatory. The Classical Numismatic Group and Atlas Coelestis gave their permission to use images of some of their items in this book.

Others to whom I owe a debt of gratitude for their help include Professor Violet Liu, Dr Michael Owen and Dr (and Father) Chris Corbally, S. J. for insightful discussions, assistance and encouragement at various stages of writing. I also appreciate the support of Chief James Waters, Chief Salvatore DiPace, Captain Daniel Magee and Lieutenant Daniel O'Keefe of the New York City Police Department.

My apologies to others who offered their time and expertise who were not mentioned here – just because you weren't mentioned, doesn't mean you are unappreciated.

And last (but certainly not least!), my wife Anh Karam, who has been unflinching in her support and who has been invariably uncomplaining about the time and attention that my writing required – often at her expense. Finally, while I acknowledge the help provided by so many, any mistakes you might find in this book are my responsibility and not theirs – they explained things properly, but that doesn't necessarily mean that I understood properly! They were kind to share their time and expertise; please don't blame them for my lack of understanding.

PHOTO ACKNOWLEDGEMENTS

The author and publishers wish to express their thanks to the following sources of illustrative material and/or permission to reproduce it. Some locations, not given in the captions for reasons of brevity, are also supplied here.

The Achilleion Palace, Corfu: p. 108; from Peter Apian, *Cosmographia Petri Apiani mathematici studiose collectus...* (Landshut, 1524): p. 43; photo Bad Buu: p. 95; with the kind permission of Barry Lawrence Ruderman Antique Maps: p. 88; Bayeux Museum, Bayeux: p. 73; photo Bibliothèque Cantonale et Universitaire, Lausanne: p. 110; from Tycho Brahe, *Astronomiae Instauratae Mechanica* (Nuremberg, 1598): p. 29; used with the permission of Henk Bril: p. 153; British Museum, London: p. 39; photo Central Library, Lucerne: p. 111; photo used with the permission of Classical Numismatic Group Inc. (www.cngcoins.com): p. 150; used with the permission of the Colonial Williamsburg Foundation: p. 152; from Franz Cumont, *Monuments figurés sur les mystères de Mithra*, vol. II (Brussels, 1896): p. 70 (top); image by D. Mitriy: p. 166; photo Ealmagro: p. 109; photo European Southern Observatory (ESO): pp. 32–3; photo European Southern Observatory New Technology Telescope: p. 69; photo European Space Agency: p. 34; photo courtesy of European Space Agency and NASA: p. 82; used with the permission of the European Space Agency (©ESA/MPS): p. 144; photo by, and used with the permission of, Mr Peter Farris: p. 66; sketch by, and used with the permission of, Mr Peter Farris: p. 67; from a manuscript of Claudius Germanicus Caesar's *Aratea* (a Latin paraphrase of Aratus's *Phainomena*) in the Universiteitsbibliotheek Leiden (Ms. Voss lat. Q79): p. 44; photo courtesy Geographicus Rare Antique Maps: p. 80; photo Jean-Pol Grandmont: p. 40; map by Orlando Hernandez, Sait Khurama, and Gretta Alexander, used by permission of Dr Alan Hildebrand and the Geological Society of Canada: p. 177; Hunan Provincial Museum, Changsha, China: pp. 68, 70 (foot);

photo E. Kolmhofer, H. Raab/Johannes-Kepler-Observatory, Linz, Austria: p. 20; photos used with the permission of Roger Langstraat: p. 65; Library of Congress, Washington, DC: pp. 44, 53, 58, 60, 80; photo Lumos3: p. 48; from *Le Magasin pittoresque*, vol. XXI (1853): p. 94; from Alain Manesson Mallet, *Description de l'universe* (Paris, 1683): p. 78; from Hiram Mattison and Elijah Burritt, *Atlas to Illustrate Burritt's Geography of the Heavens* (New York, 1856): p. 80; from Pierre-Louis Moreau de Maupertuis, *La Lettre sur la comète de Maupertuis* (Paris, 1742): p. 110; Musée Condé, Chantilly (MS 65, f.142v): p. 74; Museo del Prado, Madrid: p. 98; Museum Rotterdam: p. 75; photo Myrabella: p. 73; NASA images: pp. 14, 126, 169, 171, 172, 181; photos NASA: pp. 6, 17, 23, 128, 129, 146, 158, 160–61, 162, 173; photos NASA (photos from the Hubble Space Telescope): pp. 100, 112, 182; NASA (photo from the Space Shuttle Columbia, 1983): p. 180; photo courtesy of NASA and the NASA Space Science Data Coordinated Archive: p. 145; National Maritime Museum, London: p. 99; image courtesy of the National Optical Astronomical Observatory, the Association of Universities for Research in Astronomy, and the National Science Foundation (NOAO/AURA/NSF): p. 57; from Isaac Newton, *Philosophiae naturalis principia mathematica . . .* (London, 1687): p. 54; Teresa Nguyen: pp. 15, 24; from Larry Niven and Jerry Pournelle, *Lucifer's Hammer* (New York, 1978): p. 116; NOAO/AURA/NSF: p. 35; from the *Ogden* (Utah) *Standard* (9 February 1910): p. 58; by Jamie Partridge and used with his permission: p. 90; developed by Professor Richard Pogge, used with his permission: p. 25; photo courtesy of Professor Richard Pogge: p. 61; photo Robin Michael Roberts: p. 135; photos Gregg Ruppel, used with his permission: p. 83; San Diego Natural History Muesum: p. 176; from the *San Francisco Call* (19 May 1910): p. 60; photo Schnobby: pp. 156–7; Scrovegni Chapel, Padua: p. 72; used with the permission of Professor Bruce Simonson: p. 174; photo courtesy of Teri Smoot and the Star Shadows Remote Observatory: p. 18; Solar and Heliospheric Observatory (SOHO): p. 148; Stift Kremsmünster, Austria: p. 49; from the *Tacoma Times* (18 May 1910): p. 81; Trinity College Cambridge Library (Ms. R.17.1, fol. 10r): p. 141; from Mark Twain, *Captain Stormfield's Trip to Heaven* (New York, 1909): pp. 122, 123; used with the kind permission of the Utica Comets: p. 124; from Jules Verne, *Hector Servadac's Voyages and Adventures Across the Solar System* (Paris, 1877): p. 120; from *The Wasp* (8 July 1881): p. 135; photo Zentralbibliothek Zürich: p. 77; photo by Eurico Zimbres and used with permission: p. 176.

INDEX